STATISTICAL QUESTIONS *from the* CLASSROOM

J. Michael Shaughnessy
Portland State University
Portland, Oregon

and

Beth Chance
California Polytechnic State University
San Luis Obispo, California

with

George Cobb, *Mount Holyoke College, South Hadley, Massachusetts*
Sandra Takis, *Arlington Public Schools, Arlington, Virginia*
Jacqueline Stewart, *Michigan State University, East Lansing, Michigan*

NATIONAL COUNCIL OF
TEACHERS OF MATHEMATICS

Copyright © 2005 by
THE NATIONAL COUNCIL OF TEACHERS OF MATHEMATICS, INC.
1906 Association Drive, Reston, VA 20191-1502
(703) 620-9840; (800) 235-7566; www.nctm.org
All rights reserved

Library of Congress Cataloging-in-Publication Data

Shaughnessy, Michael, 1946-
    Statistical questions from the classroom / J. Michael Shaughnessy and
Beth Chance, with George Cobb, Sandra Takis, Jacqueline Stewart.
        p. cm.
    Summary: "Consists of eleven short discussions of frequently asked ques-
tions about statistics raised by students and by classroom teachers. Offers
teachers of statistics some insight and support in understanding these issues
and explaining these ideas to their own students"--Provided by publisher.
    Includes bibliographical references.
    ISBN 0-87353-582-0
1. Mathematical statistics--Examinations, questions, etc. 2. Statistics--
Study and teaching. I. Chance, Beth L. II. Title.
    QA276.2.S53 2005
    519.5'071--dc22
                                2005013932

Printed in the United States of America

Table of Contents

Preface

This little book consists of eleven short discussions of some of the most frequently asked questions about statistics. In our experience, both in teaching statistics ourselves and in working with classroom teachers of statistics, these questions are consistently raised by statistics students and by classroom teachers alike. Some questions such as "What is the difference between a sample and a sampling distribution?" or "What is a margin of error?" or "What is a p-value?" involve major concepts in statistics. Other questions such as "Why are deviations squared?" or "Why do we divide by $n - 1$ instead of n?" deal with some of the more technical aspects of the mathematics in statistical theory. In either instance, these questions about statistics are among the ones that have challenged us the most as teachers of statistics when we try to explain things to our students. Our primary goal is to offer teachers of statistics some quick insight and support in understanding these issues and explaining these ideas to their own students.

Most certainly there are many other questions that we could have included in this volume. The eleven questions chosen for the book made the final cut and were unanimously chosen by an NCTM editorial panel that included George Cobb, Sandra Takis, Jacqueline Stewart, and John McConnell. We thank them for their thoughtful planning in launching this book. We also wish to thank George, Sandra, and Jacqueline for helping with the initial drafts of some of the chapters and Jerry Grossman of the NCTM Educational Materials Committee for a fresh reading of the manuscript.

In addressing these questions, we have peppered the book with examples and with visual representations of the ideas. We hope these will prove useful to teachers of statistics and to their students. Happy reading and reflecting!

Mike Shaughnessy
Beth Chance

1

What Can We Learn from the Shape of the Data?

Graphical representations of data can help us to tease out the story that emerges from the context behind the data. Of course, there are also times when graphical representations can mask elements of a story as well, depending on the choice of representations. Principally we learn aspects of the story behind the data from such graphical aspects as shape, center, and spread. Also, if we plot the data versus time, we can often see illuminating trends as well.

For a quantitative variable, the shape, center, and spread can be highlighted by graphical representations such as stem-and-leaf plots, dot plots, or histograms, since either the individual data values themselves are preserved (such as in stem-and-leaf plots) or intervals of data values are preserved (such as in histograms). However, box plots can mask information, particularly about shape. Students need to be consistently encouraged (1) to relate their discussions of the data to the context in which it was measured in order to give meaning to the shape of the data, and (2) to dig behind particular representations of the data.

COMPARING DATASETS

Example 1: Measurements of Hat Size

For example, consider the data in table 1.1, which were gathered by ninth-grade students as they each measured the head circumference of one of their classmates.

TABLE 1.1
Repeated Measurements of the Head Circumference of One Student in Centimeters

Trial 1	52	52.5	52.5	52.5	53	53	53.5	53.5	54	54	54	54
	54	54	54.5	54.5	54.5	54.5	55.5	55.5	56	56	56	56
Trial 2	52	53	53	53	53	53	53.5	53.5	53.5	53.5	53.5	53.5
	53.5	54	54	54	54	54.5	55	55	55	55	55	56

In this class, for Trial 1, twenty-four students all measured the "hat size" of the same classmate (the same person throughout) to the nearest half-centimeter. According to students' suggestions, the data were graphed in both box plots and dot plots. The measurements were done on the first trial without any planning or discussion about uniformity, each student doing it his or her own way. After graphing the results, the students were not very happy about the amount of variability they obtained. They believed there should be "one correct hat size." After a discussion about ways to control for error in measurement and about variability, the twenty-four students again gathered a set of repeated head-size measurements for the same student. The box plots and dot plots for the head measurement data appear in figure 1.1.

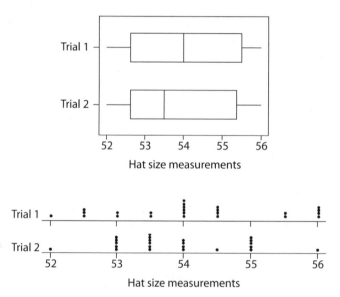

Fig. 1.1. Measurements without (Trial 1) and with (Trial 2) quality-control steps

Notice, as our students did, that the two box plots are nearly identical in size (spread) and placement (center). However, in the dot plots, the second distribution of repeated measurements is much "tighter" than the first. The first dot plot has two-thirds of the head measurements from 53.0 to 55.5 cm, whereas the second dot plot has two-thirds of the repeated head measurements from 53.5 to 54.5 cm. The quality-control measures that the students implemented helped to control some of the variability in their measurements. Displaying these data using box plots alone would mask the effect of the quality-control steps that the students took for their second round of measurements.

INTERPRETING GRAPHS—STORIES IN THE DATA

Example 2: Island Eday Populations

What can we learn about the population of the island of Eday from the shape of the graph in figure 1.2?

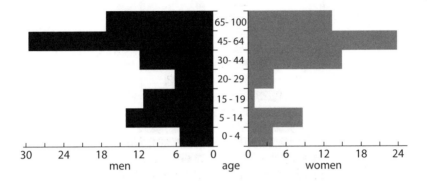

Population-pyramid of the Island Eday (1970)

Fig. 1.2. Frequencies for different age groups on an island in northern Scotland (reproduced from *Data Visualization*, by de Lange et al. 1992)

Here are a few observations, and you may have others. There are not very many young children on this island. More than half the population is at least forty-five years old. There are telling "dips" in the graph for both young men and young women, early twenties for men and late teens for women. This island seems to be "dying." The young people may be leaving for mainland Scotland or other places, perhaps because there are not enough jobs on the island. The birth rate on the island may be declining, since young people seem to be leaving and mostly older people of non-child-rearing age remain on the island. Also, for some reason, there are more men than women on the island. The shape of this data set raises a number of questions about the future viability of living on this island.

Example 3: Populations in Afghanistan

Figure 1.3 is a census pyramid for Afghanistan from 2001. What trends do we notice in this population graph? What are some possible reasons for these trends? Why might the shape of this graph cause concern in the context of radicalized groups that emerged in 2001?

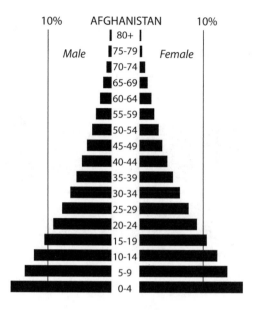

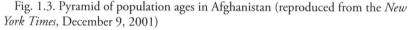

Fig. 1.3. Pyramid of population ages in Afghanistan (reproduced from the *New York Times*, December 9, 2001)

The majority of the population of Afghanistan is quite young; more than two-thirds of the population is under thirty years of age! If all ages are being counted with the same accuracy, the graph suggests that people do not live long in Afghanistan. Several decades of wars, poor health care, and perhaps insufficient food have taken their toll on Afghanistan. It appears to be a very hard life there. With regard to radicalized groups, about 50 percent of the population is between five and twenty-five years of age, a time when young people are quite impressionable and the influence of elders can have the most impact. However, this culture may have a fast "turnover" in attitudes, since there aren't many "elders" around to temper the radical beliefs with a more balanced historical approach.

FIRST IMPRESSIONS MIGHT "SAY TOO MUCH"

Example 4: HIV Infections

There are times when first impressions from a graph can actually say too much, so that the graph misleads. The graphs in figure 1.4 show the infection rates of several diseases and conditions for men and for women during the years 1982 to 1994. Students need experiences that enable them to see that the details of a graph, such as scaling, can obscure graphical meanings.

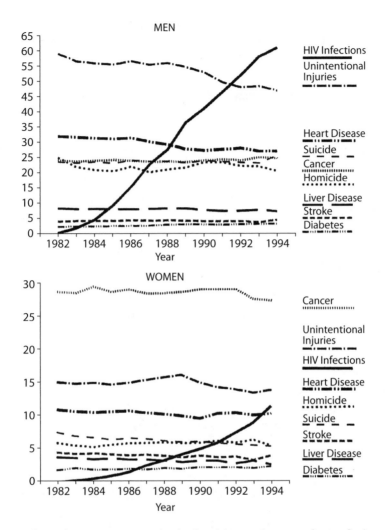

Fig. 1.4. Incidence per 100,000 (male, female) over the years of some fatal or potentially fatal events (reproduced from *The Visual Display of Quantitative Information*, by Edward Tufte, 1983)

At first glance, the cancer infection rate appears much higher for women than for men. It is somewhat higher, though not as dramatic as suggested by these two graphs, originally appearing side by side, with different scales. The cancer rate for women is about 28 or 29 per 100,000 over those years, and relatively constant, which is not really that much different from the 25 or 26 per 100,000 indicated for men. Perhaps a more important issue in these graphs is the comparison of the *increase in the rates* of infection for

HIV. Over the twelve years, the increase in the infection rate is about five times higher for men than for women (the slope of the male graph is nearly five times the slope of the female graph for HIV).

FIRST IMPRESSIONS MIGHT "SAY TOO LITTLE"

Example 5: The 1970 Draft Lottery

At times there are patterns in the data when at first glance there appears to be total chaos. Consider the graph in figure 1.5, which plots the draft numbers against birthdays for the 1970 draft lottery that was held at the height of the Vietnam War. The dot at the lower right corner of the graph, (365,3), indicates that the 365th day of the year (December 31) was pulled third in the draft order. All eligible males with birthdays on December 31 would therefore be drafted third, and similarly for all the other dots. If the lottery were totally fair, that is, "random," we would expect about the same number of dots in each month (roughly a thirty-day period) to appear in any horizontal or vertical strip of the graph. We'd expect a complete "shotgun scattering" of the dots, so to speak: we'd expect no relationship.

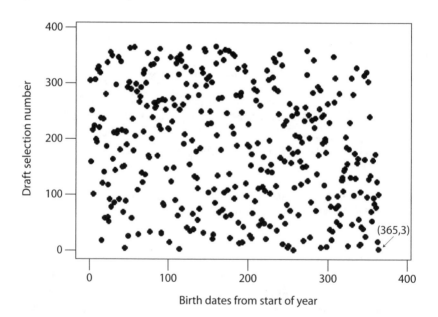

Fig. 1.5. 1970 draft lottery data

However, when the median points are traced (see fig. 1.6) for each month, we do not obtain a roughly horizontal line; rather, there appears to be a downward trend in the medians, suggesting that the later in the year one's birthday occurred, the more likely one was to have been assigned a lower draft number. More evidence for this downward trend is shown in the family of monthly box plots for the draft data, also depicted in figure 1.6.

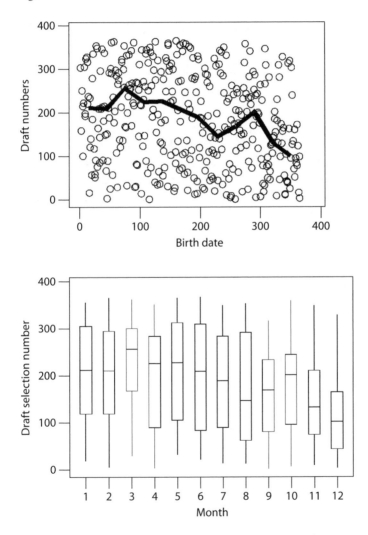

Fig. 1.6. Visual trends in 1970 draft lottery data (based on data contained in *Introduction to Basic Statistics*, by Moore and McCabe, 1989)

What at first appeared to be totally random was in fact not. In the 1970 lottery, the days of the year were written on Ping-Pong balls and stacked in a box one month on top of another, with December on the bottom and January on the top. Prior to the drawing, the box was inverted and dumped into another container, and the balls were supposedly "mixed up" in the new container before being drawn. But the mixing clearly wasn't done well, since the dates from later in the year were more likely to be drawn sooner. Buried within what initially seems to be chaotic data is actually an important piece of telltale evidence—the 1970 lottery was not fair!

CONCLUSION

An important lesson is embedded in our examples and discussion in this chapter: visual representations of data can mask, or highlight, important pieces of information. We should always take a critical look at claims made, or not made, from visual representations of data, and perhaps we should compare and contrast several displays of the data to make sure that we are not missing, or masking, important information. We also need to constantly consider and connect to the context within which the data were gathered.

2

What Do *r* and *r²* Tell Us?

When examining the relationship between two quantitative variables, we often hope to use information from one variable to predict values for the other. Although we may begin by examining graphical displays of the relationship, it is also informative to have a numerical measure of how strong the association is—and even how reliable such predictions will be.

THE CORRELATION COEFFICIENT

Example 1: Housing Prices in Bakersfield

The following data are the prices in thousands of dollars for a sample of $n = 25$ homes in Bakersfield, California. A dot plot for these data is given in figure 2.1.

95.0	133.5	117.0	65.0	98.5
120.0	132.0	162.0	162.0	130.0
89.0	185.0	92.0	205.0	106.0
155.0	126.0	244.5	154.5	169.0
135.0	170.0	165.0	200.0	186.0

Fig. 2.1. Price of homes (in thousands) in Bakersfield, California, 2003

Clearly there is a fair bit of variation in these prices (min $65,000, max $244,500). If we were to select another house at random and were asked to predict its price, we might use the mean of the sample above, $\bar{x} = 143.88$, or $143,880, as our prediction. The standard deviation of the sample, 42.36, or $42,360, gives us some indication of how far a typical observation could deviate from this prediction. Of course, if we're saving up to buy a house, we'd like to be able to predict a house price to within an amount closer than $42,360!

To improve our prediction of a house's sale price, we might take into account additional information. For example, the size of the house probably tells us a lot about why some houses are more expensive than others. The scatterplot in figure 2.2 displays the relationship between the price of the house and the square footage for the sample above.

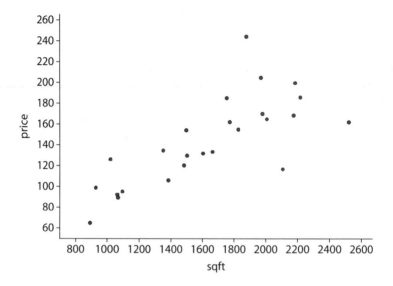

Fig. 2.2. Price in thousands vs. square footage of Bakersfield homes

This scatterplot displays a fairly strong, positive, linear relationship between these two variables: smaller houses cost less than larger houses. We can even construct a numerical measure of the strength of this association by calculating Pearson's correlation coefficient r. In most textbooks, the calculation of the correlation coefficient is shown as the sum of the products of the standardized x-values and the standardized y-values, divided by $n - 1$, namely,

$$r = \frac{\sum_{i=1}^{n} \left(\frac{x_i - \bar{x}}{s_x} \right) \left(\frac{y_i - \bar{y}}{s_y} \right)}{n - 1}. \tag{1}$$

When we interpret this formula, it is helpful to place the $\bar{x}$ and $\bar{y}$ lines onto the scatterplot, as in figure 2.3.

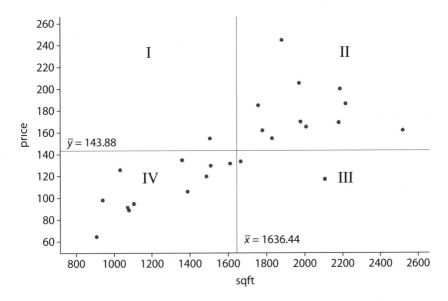

Fig. 2.3. Price and square footage of Bakersfield homes with lines at $\bar{x}$ and $\bar{y}$

Notice in figure 2.3 that there are a lot of houses in regions II and IV. When a house has a square footage, x_i, above $\bar{x}$ = 1636.44, it tends to also have a price, y_i, above $\bar{y}$ = 143.88. Such houses (region II) result in a positive component in the sum above. Houses in region IV have all $x_i <$ $\bar{x}$ and $y_i < \bar{y}$, so each of the products in formula (1) for the correlation coefficient will be positive. Since on the one hand most of the points in figure 2.3 make positive contributions to r, the overall value of r is positive. Thus, positive associations result in positive values for r. On the other hand, negative correlations result when x and y generally vary in opposite directions from their respective means, as in figure 2.4. If one of the standardized scores in formula (1) is negative and one positive, the result will be a negative product. Since most of the points in figure 2.4 are in regions I and III, the value of r is negative. A correlation close to zero occurs when there are offsetting positive and negative products, resulting in a sum close to zero, and no apparent association trend in the data, such as in figure 2.5.

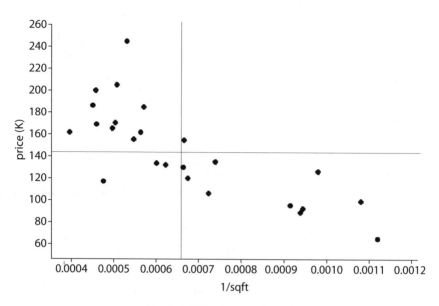

Fig. 2.4. Negative correlation

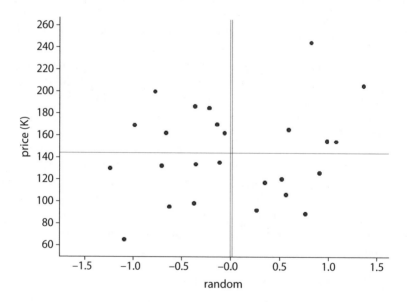

Fig. 2.5. Correlation near zero

Using a computer or calculator to compute *r*, we find *r* = .751 for the correlation between housing prices and square footage in our data from figure 2.1. The largest values that *r* can take in formula (1) is +1 for positive associations and −1 for negative associations. (For example, if you use the same variable for *x* and *y* in formula (1), you will see *r* = 1 when the correlation is perfect.) A coefficient of .751 tells us there is a pretty strong relationship between the two variables of house size and price.

THE COEFFICIENT OF DETERMINATION

Knowing the size of the house, then, would be helpful to us in predicting the cost of the house. Instead of using the overall average price (143.88), we would instead use the regression line to make such predictions (see fig. 2.6).

$$\hat{price} = 30.1 + .0694sqft$$

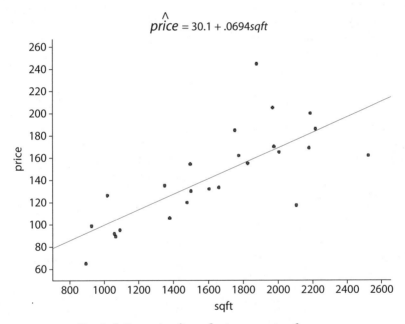

Fig. 2.6. Regression line of price on square footage

Clearly, we still won't be able to predict the housing prices exactly, but at least we are able to predict more accurately than when we just used the mean of the house prices. To measure how well our predictions do, we will look at the size of the "prediction errors." The "residual" is defined as the difference between the actual price and the predicted price. For example, for the first home, which cost $95,000, the residual from using the sample mean as a predictor is 95 − 143.88 = −48.88, or $48,880 below the actual price. But since this house has only 1092 square feet, we are less surprised at its low cost. The regression line would have predicted a cost of $\hat{y}_1| = 105.88$ ($105,880) for 1092 square feet, a residual of $95 - 105.88 = -10.88$, or only $10,880 below the actual price, much closer to the actual house price than just using the mean house price as our predictor!

This is just one house, and we want to "total" this prediction error over all twenty-five houses. We will square these differences so that the positive and negative values don't cancel each other out. The sum of these squared differences of the house prices from the mean price is

$$\sum_{i=1}^{25}\left(y_i - \bar{y}\right)^2 = 43069. \qquad (2)$$

(Notice that dividing Equation 2 by $n - 1$ and taking the square root gives us our standard deviation of $42,360, calculated at the beginning of this example). However, the sum of the squared residuals from the regression line values is

$$\sum_{i=1}^{25}\left(y_i - \hat{y}_i\right)^2 = 18793,$$

which tells us that overall the "error" is much smaller when we use the regression line to predict housing prices instead of using only the average house price as a predictor. In fact, the unexplained error (by the regression line) is only a fraction of the original error, $18793 / 43069 = 0.436$ (or 43.6%). In other words, $100\% - 43.6\% = 56.4\%$ of the variation in housing prices has been explained by the regression on square footage.

This calculation is often referred to as the "coefficient of determination." You'll note that if we took the correlation coefficient ($r = .751$) and squared it, we would obtain the same proportion, $.564$, or 56.4%. The notation used for this is r^2, and the interpretation of r^2 is the "percentage of variability in the y variable explained by the regression line with the x variable." This is distinct from the interpretation of r itself, which is a measure of the strength of association of the two variables. When $r = +1$ or -1, then $r^2 = 100\%$. This corresponds to perfect predictions, with no error, and data that fall perfectly on a straight line.

Example 1 (cont.)

Another variable that we could explore in the house example is "number of bathrooms" in the homes. We see in figure 2.7 that, not unexpectedly, there is also a positive linear relationship between price and the number of bathrooms; however, the relationship is not as strong as it was for square footage.

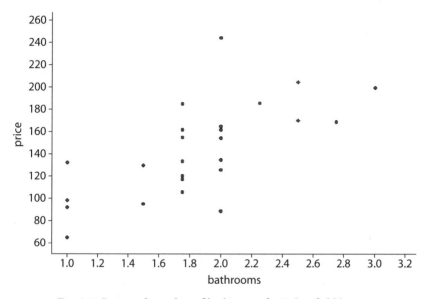

Fig. 2.7. Price and number of bathrooms for Bakersfield homes

The correlation coefficient for price and the number of bathrooms is $r = .672$, and the coefficient of determination is therefore 45.2%, indicating that 45.2% of the variation in housing prices in this sample is explained by the regression on the number of bathrooms. Thus, square footage is more helpful to us in predicting the price of a home in Bakersfield than the number of bathrooms.

Example 2: Raleigh Temperatures (Rossman and Chance 2001)

The scatterplot in figure 2.8 displays the average monthly temperatures in Raleigh, N.C., vs. the number of the month (January = 1, February = 2, and so on). Here are some possible questions we might pursue with these data.

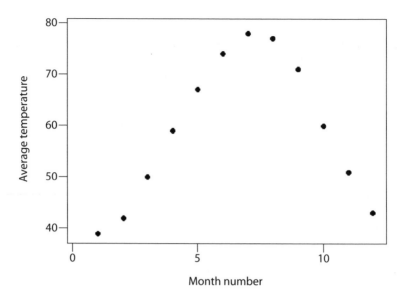

Fig. 2.8. Average Raleigh temperatures by month

- *Does there appear to be any relationship between temperature and the month in Raleigh?*
- *Does this association appear to be strong or weak?*
- *What do you think the value of r will be?*

According to the graph in figure 2.8, there is clearly a relationship between temperature and the months, but it is not a linear relationship. The average monthly temperature is highest in midyear in Raleigh (June, July, and August, or months 6, 7, and 8). However, even though there is a relationship, the correlation coefficient value r turns out to be only .257, which indicates a fairly weak positive association between the two variables and suggests that a linear model would account for only about $(.257)^2 =$ .065, or about 6%, of the variance in temperature. Thus the increasing month number is not a very good predictor of temperature. It is important to remember that r measures the strength of the *linear* association between two variables. More complicated types of relationships (e.g., quadratic) can go undetected by r. (From the shape of this graph, we might consider trying to model the data with a quadratic, rather than a regression, line.)

Example 3: Love and Marriage (Scheaffer et al. 2004)

The scatterplot in figure 2.9 displays the relationship between marriage and divorce rates (per 1000 people per year) for fourteen countries, the

United States and thirteen European countries. Here are some possible questions we might pursue with these data.

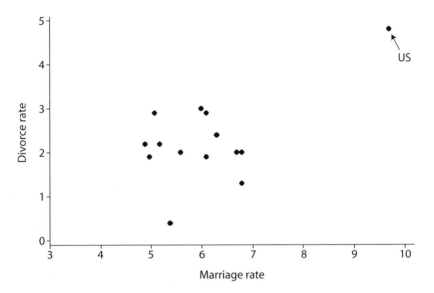

Fig. 2.9. Divorce rate and marriage rate for fourteen countries

- *Does there appear to be a relationship?*
- *Is there a linear relationship?*
- *Is it strong or weak?*
- *What do you think the correlation coefficient will be?*

The correlation coefficient for these data is actually $r = .597$, indicating a moderate positive linear relationship. This is actually a bit surprising, since figure 2.9 doesn't "look" very linear; rather, it appears to be more of an unrelated cluster of points. The least squares regression line shown in figure 2.10 has this equation

$$\hat{divorce} = -.66 + .481\,marriage,$$

and the coefficient of determination is $r^2 = 35.6\%$, indicating that 35.6% of the variation in divorce rates can be explained by the regression with marriage rates. Thus, knowing the country's marriage rate appears to be of some help in predicting the country's divorce rate. However, it is important to notice that one country, the United States, stands out in both marriage rate and divorce rate. If the United States is removed from this data set, as in figure 2.11, the equation becomes

$$\hat{divorce} = 2.41 - .056 marriage,$$

and there would not be much relationship at all between these two variables for just the European countries, since $r = -.055$. Thus, the United States

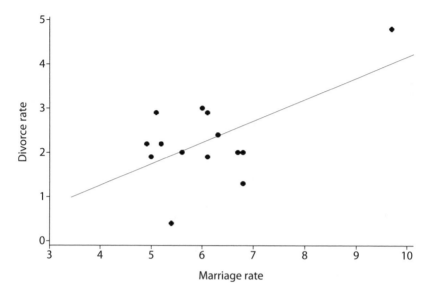

Fig. 2.10. Regression line for marriage rate and divorce rate

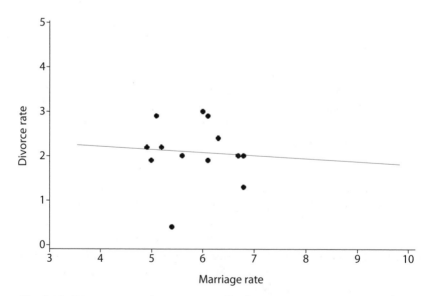

Fig. 2.11. Divorce rate and marriage rate for thirteen countries, removing the United States

exerted a strong influence on both the regression equation and our measures of the strength of the relationship. Its extreme position actually pulled the regression line very near to it (resulting in a small residual) and made the linear relationship seem much stronger. In fact, if instead the United States had a marriage rate of 9.7 but a divorce rate of 0.8, we would obtain the regression line in figure 2.12. Then we would have had evidence of a moderately strong negative relationship ($r = -.41$) between the marriage rate and the divorce rate!

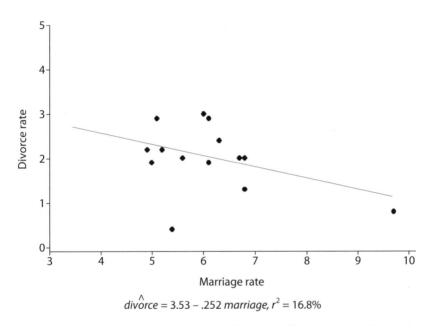

$$\hat{divorce} = 3.53 - .252\ marriage,\ r^2 = 16.8\%$$

Fig. 2.12. Divorce rate and marriage rate placing the United States at (9.7, 0.8)

This example points out that both r and r^2 are highly sensitive to influential observations or groups (typically those occurring at extreme x values). The moral of this story is that it is important to look carefully at graphs and not just blindly pick the model with the highest r^2 value.

CONCLUDING REMARKS

It is important to remember that r and r^2 measure only the amount of *linear* association. If the relationship between two variables is nonlinear, as in figure 2.7 above, then r and r^2 should not be reported. Sometimes transformations can be performed to obtain other variables that are in fact linearly related; alternatively, nonlinear models can be applied instead. The key is to carefully examine scatterplots and visual displays of the data first. There could also be clusters or subgroupings of the data that affect numerical summaries in ways that would be misleading if the data were examined without those clusters or subgroups. In general, it is important to realize that r and r^2 are not *resistant* measures, that is, they can be highly influenced by outliers, as they were in Example 3 above. Generally, observations with x values far from $\bar{x}$ tend to be more influential on the value of r.

3

Why Are Deviations Squared?

In statistics, whenever there are measurements involving "distances from," such as the distance of a data value from the mean or the distance of a point from the regression line, you have probably noticed that such measurements invariably involve "squares of differences," or "sums of squared differences." A question that often pops up in introductory statistics classes is the following: Why are deviations squared? Why isn't the sum of the absolute differences used rather than the sums of squared differences?

You may have heard explanations such as "Things work out more cleanly if we use squared deviations instead of absolute deviations" or "The calculus and the analysis are nicer for squared deviations." A bit of discussion around an example or two may help to explain why squared deviations are better than absolute deviations.

THE CASE OF UNIVARIATE DATA

Figure 3.1 contains some data about the protein content of some fast-food sandwiches. Suppose that we wanted a "representative" value for this entire data set, one that would give a good estimate for the protein of any of these sandwiches. That value would need to be "as close as possible" to all the protein values in the data set in order to minimize the error in our estimate. If we let x stand for this minimizing value, then we would want the distance from x to each of the points in figure 3.1 to be as small as possible.

The distance from x to any of the values in figure 3.1 is called the *residual* for that value. For example, $(25 - x)$ is the residual distance for a Big Mac. A residual could be positive or negative, so the absolute deviation, $|25 - x|$, furnishes a positive distance of the value 25 from x. In order to minimize all the residuals for figure 3.1 simultaneously, one way might be to try to locate a value of x so that the sum of the absolute deviations from x would be a minimum, that is, so that $|x - 46| + |x - 25| + |x - 14| + \ldots + |x - 31|$ is as small as possible.

Sandwich	Protein (grams)
Bacon Ultimate Cheeseburger—Jack in the Box	46
Big Mac—McDonald's	25
Hamburger—Carl's Jr	14
Hamburger—McDonald's	12
Jumbo Jack—Jack in the Box	20
Single—Wendy's	24
Wendy's Big Bacon Classic	34
Whopper—Burger King	31

Fig. 3.1. Protein in grams in some fast-food sandwiches (reproduced from *Nutritional Facts* on company Web sites, October 2004)

Minimizing Two Residuals

A simpler example of this would be for us to locate x so that we could minimize just two of the residuals simultaneously, say, $|x - 25| + |x - 14|$. Figure 3.2 shows the graph of the function $f(x) = |x - 25| + |x - 14|$. Notice that the "corners" in the graph in figure 3.2 occur *at* the data values 25 and 14. Also notice that this function does *not* have a unique minimum value, since the minimum is 11 for the entire interval from $x = 14$ to $x = 25$. Sums of absolute value functions have flat spots and corners. These functions are not differentiable everywhere in their domain, nor do they necessarily have unique local minimum values. That means they are neither very practical for finding "best" minimizing values nor convenient for using calculus to help find a value of x that would minimize the sum of the absolute deviations.

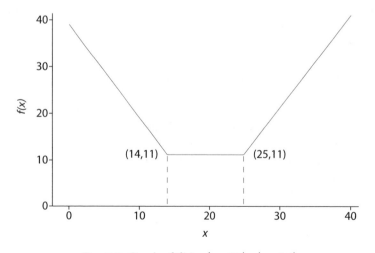

Fig. 3.2. Graph of $f(x) = |x - 25| + |x - 14|$

However, if we attempt to minimize functions that are sums of *squared deviations*, whose graphs are smooth, we will always be able to find a unique minimum value. For example, suppose we consider the function $g(x) = (x - 25)^2 + (x - 14)^2$. The graph of this function is given in figure 3.3.

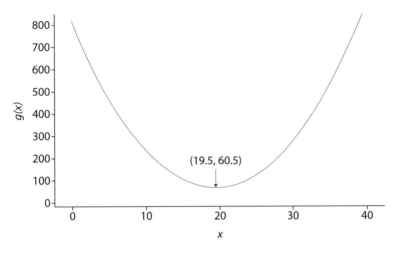

Fig. 3.3. Graph of $g(x) = (x - 25)^2 + (x - 14)^2$

This function has a *unique* minimum value. In fact, we can show that the exact minimum occurs right at $x = 19.5$. The function $g(x)$ can be rewritten as $(x - 25)^2 + (x - 14)^2 = 2x^2 - 78x + 821$, which is a parabola, a second-degree polynomial of the form $y = ax^2 + bx + c$. Parabolas have a unique minimum (or maximum) value at $x = -b/2a$, in this instance at $x = -(-78)/4 = 19.5$. Notice that the value $x = 19.5$ is actually the mean of the points 25 and 14. So the sum of the squared residuals for these two values of protein, 25 and 14 grams, is minimized by their mean value.

Graphs of the functions for the absolute deviations and the squared deviations for the full data set are displayed in figures 3.4 and 3.5, respectively.

The function $g_1(x)$ is the sum of parabolas, which always yields another parabola of the form $ax^2 + bx + c$. This parabola has a unique minimum at $x = 25.75$, which happens to be the mean of the eight data values. If calculus is used, it can be shown that the mean will minimize the squared deviations for a data set of any size.

However, the function $f_1(x)$ is the sum of absolute value functions, and this sum does not lead to another absolute value function of the form $k(x) = |x \pm c|$ for some real number c. Rather, it is a piecewise linear function with corners corresponding to each of the eight data values. Although quadratic functions are closed under addition, absolute value functions are not. Notice that the minimum value of $f_1(x)$ occurs anywhere along the flat spot between

$x = 24$ and $x = 25$ (see fig. 3.4). If we sort the eight data values from smallest to largest, these two values correspond to the fourth and fifth data values. When there is an even number of data values, the graph of the sum of the absolute deviations will always have such a flat spot between the middle two data values (assuming they are distinct). Thus, any value between the two middle values minimizes the sum of the absolute residuals. In particular, this function is minimized by the *median* value of the data set. The median could be reported as any value between the two middle values but is commonly reported as the midpoint of the flat segment, in this example, 24.5.

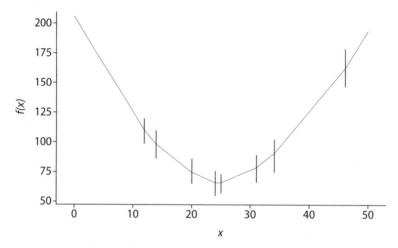

Fig. 3.4. Graph of $f_1(x) = |x - 46| + |x - 25| + |x - 14| + |x - 12| + |x - 20| + |x - 24| + |x - 34| + |x - 31|$

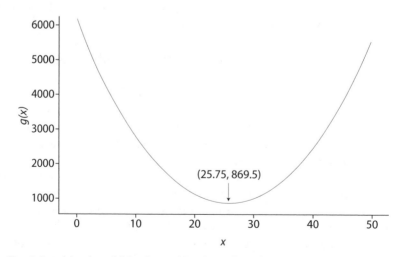

Fig. 3.5. $g_1(x) = (x - 46)^2 + (x - 25)^2 + (x - 14)^2 + (x - 12)^2 + (x - 20)^2 + (x - 24)^2 + (x - 34)^2 + (x - 31)^2$

If we had an odd number of data values (e.g., deleting the Bacon Ulti-mate Cheeseburger from the list in fig. 3.1), the function $f_2(x)$ will have a unique minimum occurring at the middle value of the data set, now 24 (see fig. 3.6). Since there is not always a unique minimizing value, the sum of the squares of deviations is preferable to the sum of absolute deviations.

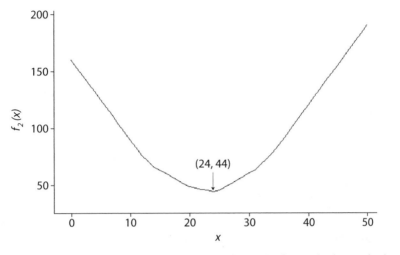

Fig. 3.6. Graph of $f_2(x) = |x - 25| + |x - 14| + |x - 12| + |x - 20| + |x - 24| + |x - 34| + |x - 31|$

THE CASE OF BIVARIATE DATA

Suppose we wish to predict how many grams of fat there might be in a fast-food sandwich if we know the amount of protein in the sandwich. How might we go about that? Consider the data set in figure 3.7, also depicted in the graph in figure 3.8.

Sandwich	Protein (grams)	Fat (grams)
Bacon Ultimate Cheeseburger—Jack in the Box	46	70.5
Big Mac—McDonald's	25	33
Hamburger—Carl's Jr	14	9
Hamburger—McDonald's	12	10
Jumbo Jack—Jack in the Box	20	28
Single—Wendy's	24	24
Wendy's Big Bacon Classic	34	29
Whopper—Burger King	31	42

Fig. 3.7. Protein grams vs. fat grams in some popular fast-food burgers

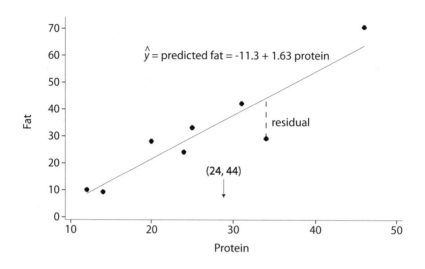

Fig. 3.8. Scatterplot and line of best fit for protein-fat data in figure 3.7

A line of best fit for this scatterplot would be located so that it would pass "as near as possible" to all the actual points on the graph, using some criteria for "as near as possible." The points (x, y) represent the "actual" fat grams for particular protein gram values, whereas the $\hat{y}$-values on a fitted line would represent the fat values "predicted" by that line for given protein value inputs. The vertical difference between the actual y-value for fat and the value for fat $\hat{y}$ predicted by the line is called the *residual* for that protein value. To find the line of best fit, we need to determine the values of a and b that minimize these residuals, $y_i - \hat{y}_i$, or $y_i - a - bx_i$. We could again consider two different functions to minimize,

$$f(a, b) = \sum_{i=1}^{n} | y_i - a - bx_i | \quad \text{and} \quad g(a, b) = \sum_{i=1}^{n} (y_i - a - bx_i)^2,$$

where these are both functions in two variables, a and b. Figures 3.9 and 3.10 display these functions using three data values, $x_1 = 46$, $x_2 = 25$, and $x_3 = 14$.

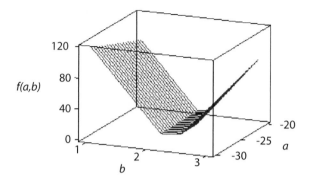

Fig. 3.9. Sum of absolute deviations

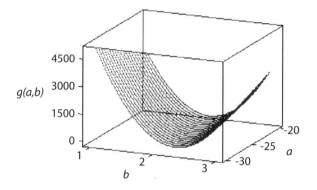

Fig. 3.10. Sum of squared deviations

As in the univariate example, the sum of the absolute deviations, $f(a, b)$, is not differentiable for all values of a and b because of its "kinks" and flat spots, but $g(a, b)$ is everywhere differentiable, producing a nice, smooth surface.

SUMMARY

The "nice" properties that "sums of squares" functions have include differentiability, closure under addition, and a unique minimum value. These properties make it possible to solve many minimizing problems in statistics, such as finding the unique line of best fit through a bivariate data set. For these reasons, the least squares criterion pervades much of statistical theory. A nice discussion of residuals and regression that is accessible to high school students can be found in *Exploring Regression* (Burrill et al. 1999).

4

What Is Independence?

The term *independence* crops up in many statistical settings—independence of events, independence of variables, and independence of trials, to name a few. But these are really special cases of the same idea: does knowing about what has already happened affect the probability of something else happening?

INDEPENDENCE OF EVENTS

The prototypical example of independence is coin tossing. If we have a fair coin, then the probability of the coin landing on its head is .5 every time, regardless of how many times we may have flipped the coin or the results of those previous outcomes. If we toss the coin three times and obtain HTH, then the probability that the next toss is heads is .5. If we toss ten times and obtain HHHHHHHHHH, then theoretically the probability that the next toss lands heads is .5. No matter what we know about the outcomes of the previous tosses, the probability of heads on the next toss remains at .5. In this example, we say the events of "heads" on different tosses are independent. Similarly, if we roll two dice, the probability that the sum will be 7 is 1/6 for every roll, regardless of the outcomes of previous rolls, and the events of "sum to 7" on different rolls are independent.

INDEPENDENCE OF VARIABLES

Let's look at a slightly more involved example.

Example 1: Gender and Color Blindness, a Case of Independence

Suppose we have data on whether or not a person is color-blind and the person's gender. Data for one such sample of 350 people are presented in the two-way table shown in figure 4.1.

	Male	Female	Total
Color-blind	10	25	35
Not color-blind	90	225	315
Total	100	250	350

Fig. 4.1. Two-way contingency table of gender and color blindness for Example 1

According to these data, if we were to randomly select one of these 350 people, the probability that person would be color-blind is 35/350 = .10. Suppose, however, that before we check for color blindness, someone tells us we picked a male. In this instance, knowing that the person is male doesn't change the probability that the person is color-blind, since according to figure 4.1, the probability is 10/100, or still .10, that this male person is color-blind. Using conditional probability notation, we would write

$$P(\text{color-blind} \mid \text{male}) = .10.$$

In this case, we say that these events, {color-blind person} and {male person}, are independent because knowing that one of these events has occurred does not change the probability of the other event. Similarly, we see that the events {female} and {color-blind} are also independent, since $P(\text{color-blind}) = P(\text{color-blind} \mid \text{female}) = .10$. In fact, you can check the table in figure 4.1 to see that the events {female} and {not color-blind} are also independent, as are the events {male} and {not color-blind}. Since these are all the possible pairings of the events, we can make a stronger statement than just that some events are independent. In this example the *variables* "whether or not the person is color-blind" and the person's "gender" are independent. Knowing the outcome of the gender variable in this sample gives us no new information about the probability that the person is color-blind.

Example 2: Gender and Color Blindness, a Case of Dependence

If, however, the data on gender and color blindness looked like the sample in figure 4.2, we would not have independence, because the probability of being color-blind changes, depending on whether a male or a female is randomly picked from the sample.

	Male	Female	Total
Color-blind	5	1	6
Not color-blind	95	249	344
Total	100	250	350

Fig. 4.2. Two-way contingency table of gender and color blindness for Example 2

If we select one of the people from the table in figure 4.2 at random, the probability that the person is color-blind is 6/350 ≈ .017. But in this instance, if we are told ahead of time that we selected a male, the probability that this male is color-blind is 5/100 = .05. Also, if we knew the person was a female, the probability that she is color-blind is only 1/250 = .004. Furthermore, P(being color-blind) ≠ P(color-blind | male), since .017 ≠ .05, and similarly P(being color-blind) ≠ P(color-blind | female), since .017 ≠ .004 either! Knowing the outcome of the gender variable would increase the chance that the person is color-blind if you learned the person was a male but decrease the chance of being color-blind if you learned the person was a female. Since the knowledge of the person's gender changes the likelihood of color blindness, gender and color blindness are in fact *not independent* variables in this sample. Rather, they are *dependent* on each other. The probability that a person is color-blind depends on (is related to) his or her gender.

INDEPENDENT OBSERVATIONS WHEN SAMPLING

Many statistical procedures require "independent observations." How do we know the observations are independent? In particular, if we are selecting a sample (without replacement) from a larger population, can we assume that the results from person to person are independent? In the sample in figure 4.2, if we randomly selected one of the 350 people, then P(color-blind) is 6/350. What about the next person? If the first person had been color-blind, then the probability that the second person is also color-blind becomes 5/349. Yet if the first person had not been color-blind, this probability would be 6/349. Since the probability that the second person is color-blind changes depending on the outcome for the first person, these are not independent observations. So technically, on the one hand when we randomly sample cases without replacement from a population, the observations are dependent. On the other hand, our intuition tells us that in a large population, the chance of being color-blind from one person to the next should be about the same. How do we address this issue? Here are two possibilities.

Sampling with Replacement

A simple way to address the difficulty above is to put the first person back! Then, the probability that the second person is also color-blind is still 6/350. This is akin to the coin-tossing example. Any time we randomly sample "with replacement" from a population or we sample from a long-run process, we can consider the outcomes of subsequent trials to be independent from one another, that is, the outcome on any one trial does not affect or change the probability of an outcome on subsequent trials.

Sampling from a Large Population

In reality, we often don't, or can't, put the people back. We often select a subset of people from a population all at once for our sample. Can we still assume the outcomes from person to person are independent? Imagine that we have an entire population of 35,000 people from a city, and 600 of them are color-blind. Then the probability that the first person we meet in the town at random is color-blind is 600/35,000 ≈ .01714 to five decimal places. Assume she is color-blind. Suppose that as we walk with her we meet a second person at random. What is the probability that the second person is color-blind as well? Since we didn't put the first person "back in," this time the probability is 599/34,999, or .01711 to five decimal places. So, for all practical purposes, with a large population (or at least a rather small sample from a large population), knowing the first person is color-blind does not appreciably change the probability that the second person is also color-blind. So if we have a very large population size N compared to our sample size, it is reasonable to treat the outcomes as independent for every member of the sample.

WHAT DO WE GAIN WHEN WE HAVE INDEPENDENCE?

What is the advantage of being able to assume that observations are independent? The short answer to this question is "easier calculations." Suppose we apply to two different graduate schools, A and B, and that we knew P(accepted into school A) = .6 and P(accepted into school B) = .7. What is the probability that we are accepted into both schools?

If we assume independence for acceptance into each of these schools, then

P(accepted into A *and* accepted into B) = $P(A \cap B)$
= $P(A) \times P(B)$
= .7 × .6 = .42,

using the "multiplication rule for *independent* events."[1] However, it is unrealistic to assume these two events are independent. The schools may be looking for similar characteristics in our transcript. We would have to first consider the probability of getting into school A, and then *given that we have made it into school A*, find the probability that we would also get into school B. In other words, we would need to know the conditional probability P(accepted into B | accepted into A). In probability notation, the general calculation for the probability that we get into school A and also get into school B is as follows:

P(accepted into A ∩ accepted into B)
$= P$(accepted into A) × P(accepted into B | accepted into A),

using the "multiplication rule for *dependent* events."[1]

Similarly, when sampling from a population, to find the probability of two color-blind people, we would need to calculate

P(first person color-blind)
× P(second person color-blind | first person color-blind)

when the individual outcomes are not independent.

Thus, having independence allows us to multiply simple probabilities and to assume that the probability of an outcome does not change because of results of previous outcomes. Computations become more difficult when events are dependent. This "convenience" is at the heart of many theoretical results in statistics.

SUMMARY

When the outcome for one event alters the probability of another event (a prior event, a future event, or a simultaneous event), those events are called *dependent*. Otherwise, they are called *independent* events. Furthermore, if none of the probabilities of any of the values of a variable changes when we know the outcome of another variable—such as in the gender and color blindness Example 1, then those *variables* are said to be *independent*. Outcomes from repeated trials of a random process with constant probability of success from trial to trial are also independent. In sampling situations where the population from which we are sampling is large compared to the size of the sample, we consider that the outcomes from one trial to another are independent for practical purposes, since probabilities of different outcomes don't change measurably from one trial to the next in this situation. In general, having independence greatly simplifies subsequent statistical calculation.

---

1. For more details on multiplication rules for *independent* and *dependent* events, see, for example, Shaughnessy et al. (2004).

5

What Is the Difference between an Experiment and an Observational Study?

Suppose that you hear that a diet rich in blueberries is related to longer life spans. If you wanted to carry out a study to explore this, should you conduct an experiment or an observational study? It depends. Experiments and observational studies are the two main kinds of comparative studies. It has been said that the distinction between them, together with the consequences of that distinction, constitute the single most important contribution that statistics has made to the progress of science.

Abstractly, the distinction is this: In an observational study, we compare groups of individuals where the group distinction is "built in"—for example, examining the life spans of those who had a diet rich in blueberries and comparing them to individuals who did not have a diet rich in blueberries. The key is that it was up to those individuals to determine which group they were in. We are just gathering information about them "after the fact." In an experiment, the individuals whom we compare start out as a single group and the researchers are responsible for dividing them into subgroups. We could find some volunteers and tell half of them to add blueberries to their diet and the other half to avoid blueberries. The researchers are imposing the diet (the *explanatory variable*) on the individuals; it is not their own choice. However, experiments cannot always be done. For example, to compare the life spans of men and women or to compare the blood pressures of retirees and teenagers, an experiment is not possible, since we can't impose gender or age on our subjects. In such situations, an observational study has to be used. But to compare the effectiveness of aspirin and a new headache medicine, or the effect of classical music versus rap on the heart rates of chickens, we can do an experiment, since we can split the subjects into groups first and then assign the conditions.

OBSERVATIONAL STUDIES: SOME EXAMPLES AND SOME DIFFICULTIES

Example 1: Smoking and lung cancer

One of the early observational studies of smoking and health compared mortality rates for three groups of men. The rates, in deaths per 1000 men, were as follows:

Nonsmokers	20.2
Cigarette smokers	20.5
Cigar and pipe smokers	35.5

If you take the numbers at face value, it appears that cigarettes are safe but pipes and cigars are dangerous. Of course in this study the investigators did not randomly assign the conditions to be compared (type of smoking), and as a result, the groups may have differed in a number of ways besides their smoking habits. The observed differences in the death rates were due mainly to another variable, the differences in average age (in years):

Nonsmokers	54.9
Cigarette smokers	50.5
Cigar and pipe smokers	65.9

Although it is clear that the mortality rate was highest among the cigar and pipe smokers, we have no way of distinguishing whether this arose from the smoking choice or from the fact that the individuals in that group tended to be older.

Example 2: Thymus surgery

In 1912, Dr. Charles Mayo, one of the two brothers for whom the Mayo Clinic is named, published an article recommending surgery to remove the thymus gland as a treatment for children with certain kinds of respiratory problems. He made this recommendation despite the fact that roughly one-third of the children he had operated on had died! The basis for his recommendation was an observational study—he compared the sizes of the thymus glands of the children he operated on to the sizes of thymus glands from autopsies of adults, and he found that the adults' glands were smaller. Mayo was comparing two groups, and he thought he was using an appropriate control group (the adults), but his two groups had not been determined using a chance device, and Mayo was misled by a *confounding variable*. What no one realized at the time was that in normal, healthy people, the thymus gets smaller as one gets older. The difference in size that Mayo had observed was due simply to the differences in age, and it had nothing to do with the respiratory problems of the children.

Moral of the story: With observational studies, we can never eliminate the possible impact of confounding variables. This prevents us from attributing cause for any differences between the groups at the end of the study to the explanatory variable of interest.

EXPERIMENTS: CREATING COMPARISON GROUPS

In an experiment, we must be very careful how we assign the treatment conditions. In order to be able to make sound comparisons, the groups must be created "at random" using a chance-based method. For example, we could flip a coin for each individual; if it lands heads, we would put the person in the blueberry group, but if it lands tails, he or she would be in the no-blueberry group. If we use a chance device to sort individuals into groups, (1) we can't tell in advance which individuals will end up in a particular group. In fact, for a completely randomized experiment, all possible allocations are equally likely. Nevertheless, (2) if our groups are large enough, we can be quite confident that the groups will "look like each other" in all important ways. For example, other variables, such as age, diet, or initial weight, should be "evened out" between the two groups. Therefore, the only substantial difference between the two groups will be the condition(s) that we assign to the groups.

Example 3: Do Democrats have longer last names?

The chart in figure 5.1 lists the presidential candidates between 1968 and 2000, their party, and the number of letters in their surname.

Election	Candidate	Party	Length
1968	Humphrey	Democrat	8
1968, 1972	Nixon	Republican	5
1972	McGovern	Democrat	8
1976, 1980	Carter	Democrat	6
1976	Ford	Republican	4
1980, 1984	Reagan	Republican	6
1984	Mondale	Democrat	7
1988	Dukakis	Democrat	7
1992, 1996	Clinton	Democrat	7
1988, 1992	Bush	Republican	4
1996	Dole	Republican	4
2000	Gore	Democrat	4
2000	Bush	Republican	4

Fig. 5.1. Name length of presidential candidates

From these data, the average length for the surnames of the Democrats is 6.71 and the average length for the surnames of the Republicans is 4.50, for a difference of 2.21 letters. But the question is, could such a difference have happened "just by chance"? The phrase *by chance* refers to a probability model. For example, what if we had the ability to randomly assign these political parties to the thirteen candidates? The frequency distribution in figure 5.2 displays the differences in the two sample averages for 1000 repetitions of a simulation that randomly assigned the Republican and Democrat distinctions among the thirteen candidates.

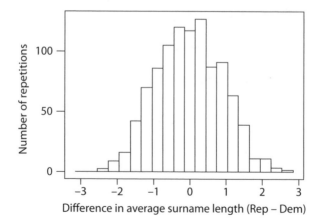

Fig. 5.2. Frequency distribution of 1000 repetitions of random assignments of candidates

This graph tells us how large we expect these differences in average length to be when the groups are created purely at random. We see that it is pretty common to see a difference in the length of surnames of up to 1 letter "just by chance," and not too surprising to see a difference of up to 1.5 letters. But the actually observed difference for our sample of more than 2 letters seems a little surprising. We have evidence from the sampling distribution in figure 5.2 that a difference this large wouldn't happen just by chance, indicating a *statistically significant* difference. (For more on statistical significance, see chapter 9, entitled "What Is a *p*-Value?")

When we randomly assign the individuals to the two groups, we see that sometimes the surnames of the Republicans are a little longer and sometimes the surnames of the Democrats are a little longer. In the long run, the surnames don't appear to be much different between the two groups under random assignment conditions. In fact, for any variable that we might measure on these individuals, in the long run, that variable should balance out between the two groups. In practice, if groups are created at random, we feel comfortable assuming there is no substantial difference between the groups. This gives us a fair comparison group. Then we assign our conditions. If we later observe a difference between the groups, knowing they should have been quite similar to begin with, we will attribute the later differences to the conditions we imposed.

In summary, randomization serves to create similar groups, so that any differences we observe can be attributable to the explanatory variable we imposed. Furthermore, probability models allow us to determine which results are typical and which are highly unusual. Asking "What is the likelihood that something would happen by chance," as in the example of the difference in name lengths above, and finding out that the probability of a difference of more than 3 letters is a rare event from natural variability alone are the basis for making a statistical inference. In the names example, we might remember that the probability model here is hypothetical and that we can't draw cause-and-effect conclusions from this observational study, but we still have a way of measuring how surprising the result is. When randomization is a part of the design itself, we can draw more powerful conclusions and attribute cause.

DRAWING CONCLUSIONS

The importance of the distinction between experiments and observational studies comes from the kinds of inferences they support. Experiments let us determine whether any differences we observe among our groups were in fact *caused* by the conditions we assigned to them; observational studies generally do not. An observational study might show that on average, women live longer than men, but it can't reveal what it is about men and women that is the cause of the difference. If a randomized experiment shows that chickens that hear Mozart have statistically significantly lower heart rates than chickens subjected to Eminem, we can conclude that the music *caused* the difference.

Example 4: Calcium and blood pressure

Investigators who examined a large sample of people noticed that blood pressure appeared to be related to calcium intake and that the relationship was strongest for black men. To confirm that it was in fact the calcium that had caused at least part of the observed effect, an experiment was needed. Twenty-one healthy black men were divided into two groups using a chance device to do the sorting. One group received a calcium supplement for twelve weeks; the control group received a placebo pill that looked identical. Neither the subjects nor the people who measured and recorded their blood pressure knew which pill a subject was taking (a double-blind study). When statistical analysis showed at the end of the study that the observed difference in average blood pressure between the two groups was too big to be due solely to the random assignment, investigators were able to conclude that the calcium supplements had caused a reduction in blood pressure.

SUMMARY

The chart in figure 5.3 summarizes the issues of control of conditions that we do or don't have and inferences about causality that we can or can't make in studies, depending on whether the study is an observational or an experimental one.

Type of study	Conditions of interest are	Inferences about cause are
Observational	Built in	Difficult or impossible
Experiments	Assigned by a chance device	Straightforward

Fig. 5.3. Comparing observational and experimental studies on conditions and inferences

What about the blueberry study we proposed at the beginning of this chapter? Should we attempt an observational study or an experimental one? It depends. If we want to be able to attribute the longer life spans to blueberries, we would have to carry out a randomized comparative experiment. This may or may not be feasible depending on the cost and time frame allotted for our research. If we relied on the information from an observational study, we might see an association between eating blueberries and longer life spans, but we would not be able to isolate the blueberry-rich diet as the cause. Observational studies definitely have their purpose; we just need to be careful that we don't forget about the potential confounding variables and overstate a causal connection between the variables. Researchers often try to "account" for all possible alternative explanations, but there could always be a factor they didn't think of. With randomized experiments, however, even potential factors not considered prior to the start of the study should be relatively balanced by the randomization process.

6

Why Use Random Samples?

The patterns in data, and what they mean, depend in important ways on how the data were produced. Poor data production can easily give misleading results. We saw in the previous chapter on experimental vs. observational studies that randomization is the key to whether or not we can draw cause-and-effect conclusions. Random samples are the key to being able to draw conclusions about a larger population from a small sample.

SAMPLING: SOME DIFFICULTIES

Example 1: Stringing students along

This example illustrates a way to see the need for random selection that can be used directly with students. Prepare a bag of strings of different lengths. (A set of about twenty-five strings with lengths between 4 and 15 inches generally works well.) Then ask students to reach into the bag without looking, mix up the strings, draw one out, measure it, record the length, and replace the string. Repeat this for a total of ten times. Then compute the average of the lengths in the sample. Now take all the strings out of the bag, measure their lengths, and compute the actual average length for the population of strings.

The dot plots in figure 6.1 are an example of student results from this activity. The top dot plot shows the lengths of the twenty-five strings, and the bottom dot plot shows the results of ten strings drawn from the bag.

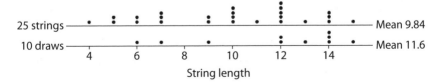

Fig. 6.1. Dot plots of the lengths of twenty-five strings and of ten randomly drawn strings

The goal is to estimate the mean string length in the entire population. However, this sampling method is biased. Longer strings are more likely to be chosen, and in fact the chance of choosing any one string is proportional to its length (which can only be computed if we know the lengths of all the strings). As a result, this procedure overrepresents the longer strings, and if we repeat this procedure many times, the averages we obtain for the samples will almost surely be larger than the population average. A biased sampling method systematically overestimates (or underestimates) the results in the population.

How Do We Obtain a Representative Sample?

A much better method for selecting a sample from a larger population is to use a chance device to decide what gets into our sample and what does not. Chancelike variability can seem paradoxical because its two main features appear to go in opposite directions: on the one hand, (1) individual outcomes are quite unpredictable; but on the other hand, (2) long-run patterns can be predicted with near certainty. When selecting a sample from a larger population, if we choose a simple random sample, (1) we can't tell in advance which individuals or items will be chosen. In fact, all possible samples (of a given size) are equally likely. Nevertheless, (2) for large enough samples, we can be all but certain that our sample will "look like the population" in that any average or proportion we compute from the sample will be close to the corresponding population value. In fact, we are also able to employ probability models that allow us to measure how far we can expect the sample result to be from the population value, known as the margin of error. (For a further discussion of this topic, see chapter 8, entitled "What Is a Margin of Error?")

Example 2: Public opinion polls

(a) In 2002, shortly after the Salt Lake City Winter Olympics ended, Mitch Romney, who had organized the local arrangements for the games, announced that he was thinking about running for governor of Massachusetts. His announcement was something of a surprise, since Romney is a Republican and so was the incumbent, Acting Governor Jane Swift. The *Boston Globe* commissioned a poll of Massachusetts Republicans and found that 75 percent preferred Romney, whereas only 12 percent preferred Swift. The newspaper stated that this sample result had a margin of error equal to 5 percent, indicating that it believed the sample result was within that distance of the population proportion favoring Romney. Not long after, Swift announced that she was withdrawing from the race.

(*b*) A few months earlier, Massachusetts State Representative Nancy Flavin sent a survey to voters in her district. Among other things, she asked if they favored using "taxpayer dollars" to pay for election campaigns. She then summarized the opinions expressed in the surveys that were returned.

The margin of error of the Romney poll determined by the *Globe* is roughly $1/\sqrt{n}$, the estimator pointed out in chapter 8 on margin of error. This number comes from a probability model that assumes that the individuals who were polled were chosen at random from the population of interest. Because the sample was chosen using a chance device, it makes sense to ask, "What is the probability that the sample might have turned out somewhat differently?" Such a probability is relevant because the mathematical model used provides a good match to the way the poll was actually conducted.

The survey by Nancy Flavin did not report a margin of error. For her poll, it would be possible to compute one, using the same formula that the *Globe* used, but the resulting number would be completely worthless. The formula still works, but the logic does not. For this poll, the mathematical model does not correspond in any reasonable way to what was actually done to produce the data. The sample was a "voluntary response" sample: the only people in the sample were those who cared enough to go to the trouble of filling out the survey and sending it back. This method tends to overestimate the unfavorable opinion and will not fairly represent the opinions of the entire population of interest. In fact, there is a second source of bias in this example: The question was worded in a way that was meant to appeal to those opposed to publicly financed elections. The wording further guarantees that the usual probability model would *not* apply. A margin of error could be computed, but since the sample was not selected randomly and the wording of the question was slanted, the computation is worse than meaningless. Sources of sampling and nonsampling bias must be considered before we generalize from the sample to the larger population.

SUMMARY OF RANDOM SAMPLING AND RANDOMIZATION

The needs for chance mechanisms discussed here are quite similar to those discussed when distinguishing observational studies from experiments (see chapter 5), and the desire to use a probability model in order to calculate "how often would such a result occur by chance" is quite the same. Whereas using "random samples" allows us to generalize from a sample to the population, employing "randomization" to assign subjects to treatment groups allows us to draw cause-and-effect conclusions. In the latter situation, we still need to consider how the subjects were selected before we generalize the results beyond the sample at hand. For example, most medical studies must rely on volunteers, and caution should be used in applying the same conclusions to the general population. Figure 6.2 summarizes when we can make inferences from a sample to a population (random samples) and when we can attribute cause to our findings (randomized trials in our design).

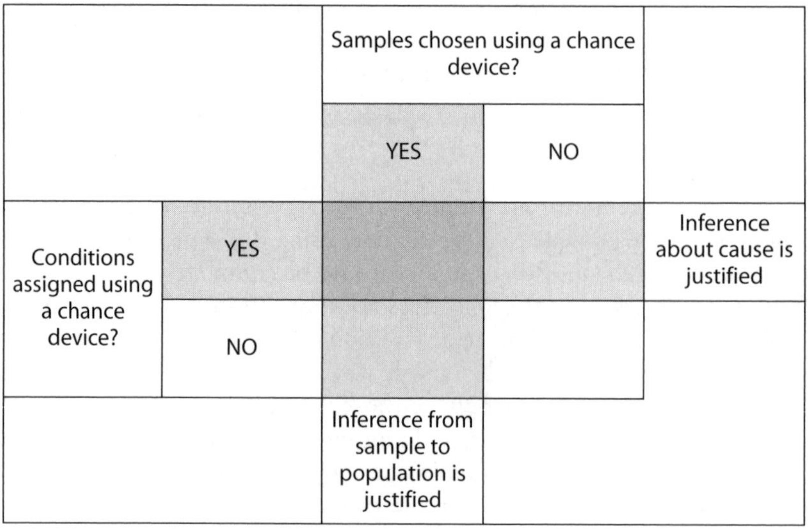

		Samples chosen using a chance device?		
		YES	NO	
Conditions assigned using a chance device?	YES			Inference about cause is justified
	NO			
		Inference from sample to population is justified		

Fig. 6.2. Inference and causation: random samples and randomized trials

7

What Are the Differences among the Distribution of a Population, the Distribution of a Single Sample, and the Sampling Distribution?

L et's begin our discussion of this question by looking at an example.

EXAMPLE 1: GRADE POINT AVERAGES—A POPULATION, SOME SAMPLES, AND A SAMPLING DISTRIBUTION

Suppose we record the grade point averages (GPAs) for all students in a school and obtain the frequency distribution for GPA shown in figure 7.1. Our population consists of all the students' GPAs in the school.

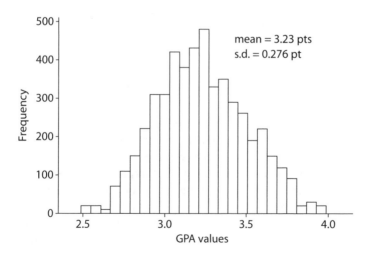

Fig. 7.1. Frequency distribution of GPAs for an entire school

If we randomly selected twenty-five students from this population, we'd expect the sample to look somewhat like the population, but the actual results will vary from one sample to another sample as illustrated in figure 7.2, which shows the results for three different samples, each of size 25.

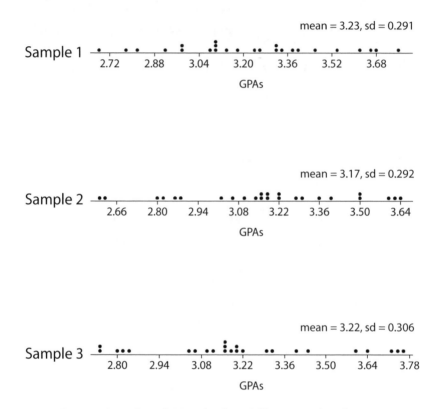

Fig. 7.2. Dot plots of GPAs for three different samples of size 25

If we calculate a *statistic* for each sample, such as the sample mean or the sample range, the value of the statistic we obtain will vary from sample to sample. Fortunately, the distribution of these sample statistics will follow a predictable pattern. The distribution of the sample statistics from all possible samples of the same size (e.g., $n = 25$) is called the *sampling distribution of the statistic*. In this example, we are interested in the sampling distribution of the sample mean. Since we can't easily draw *all* possible samples of size 25, we can approximate the exact sampling distribution by simulating an "empirical" sampling distribution. The graph in figure 7.3 shows results from drawing 400 different random samples, each consisting of 25 students, from the population above, and calculating the sample mean for each sample.

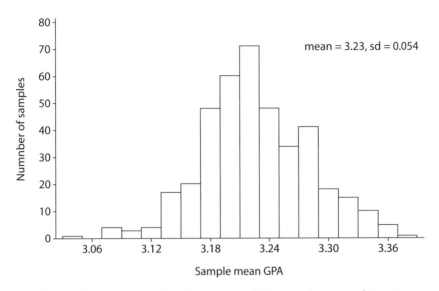

Fig. 7.3. Empirical sampling distribution of 400 sample means of size 25

Studying sampling distributions enables us to understand how summary statistics behave. For example, the sampling distribution of the sample means in figure 7.3 has a similar center to, but a much smaller spread than, the population distribution shown in figure 7.1 and the distributions of individual samples of 25 students shown in figure 7.2.

Using our knowledge about the behavior of the sampling distribution, we can draw conclusions or make inferences about a population using only data from a sample. However, many students are introduced to the process of statistical inference without having a sound understanding of sampling distributions. The primary source of their confusion is often the distinction between the distribution of a single sample and the sampling distribution of a sample statistic. To an expert, the difference between a sampling distribution and the distribution of a single sample may seem obvious. However, the terms *sampling distribution* and *the distribution of one sample* are often confused by beginning statistics students. Sometimes it may be difficult to distinguish whether it is only the terms that are being confused or if it is the concept of a sampling distribution.

For instance, suppose you pose the question to your students, "What is the mean of the sampling distribution of sample means?" A common student response is, "Its mean should be a good approximation of the mean of the population as long as the sample is randomly selected." From this statement, it is unclear whether the student has understood the question because he or she talks about one sample being randomly selected; the statement made is correct if the student is talking about an individual sample. The student has failed to clarify that we are (1) taking the means from numerous samples and looking at the long-term properties of this process and (2) asking about the mean of the sample means (which should also be close to the population mean). Here are some questions that can provide more insight into students' understanding of sampling distributions.

- What are the elements of the sampling distribution? What are the elements of a sample?

Dot plots can be helpful for asking students these questions. For example, figure 7.4 is a dot plot of an individual sample (sample 2 from fig. 7.2), and figure 7.5 is a dot plot of the empirical sampling distribution shown in figure 7.3. It is important to assign suitable labels to the horizontal axis of each of these graphs to help distinguish them.

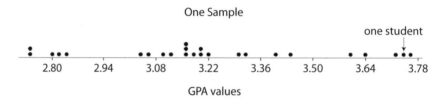

Fig. 7.4. Dot plot of an individual sample of 25 students' GPA

In the graph shown in figure 7.4, each dot represents the GPA of an individual student. In the graph in figure 7.5, each dot represents a sample, and we have recorded the average GPA for the twenty-five students in that sample.

- How is the standard deviation of a single sample related to the standard deviation of the population?

Notice in the example in figure 7.1 that the population standard deviation was equal to 0.276 point. In the three samples shown above, the sample standard deviation changes from sample to sample, but it was always around 0.3, similar to the population value.

- How is the standard deviation of the sampling distribution related to the standard deviation of the population?

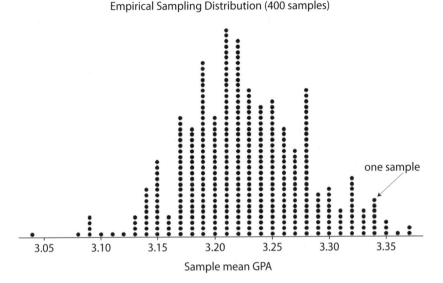

Fig. 7.5. Dot plot of an empirical sampling distribution of 400 sample mean GPAs for samples of size 25

For example, how does the standard deviation of the average GPAs for different samples (fig. 7.3) differ from the standard deviation of the GPAs of all the individual students (fig.7.1)? In figure 7.1, the standard deviation for the population of GPA values was 0.276 point, whereas in figure 7.3 for the empirical sampling distribution the standard deviation was 0.053 point. These observations should agree with students' intuition— that "averages" vary less than individuals.

• How is the shape of a single sample related to the shape of the population distribution?

Although it is difficult to tell the shape on the basis of only twenty-five observations, students should believe that the shape of the sample will also be similar to the shape of the population. That is, the center, spread, and shape of any individual sample should mimic those of the population, but there may be considerable variability from sample to sample.

• How is the shape of the sampling distribution related to the shape of the population distribution?

This last question is a little hard for students to grasp, and they probably need to see lots of examples that illustrate that the shape of the sampling distribution does not need to be the same as the shape of the population (see Example 2 below) and, in fact, is less so as the sample size increases.

EXAMPLE 2: THE SAMPLING DISTRIBUTION OF THE MEANS FOR A BINOMIAL POPULATION

It is particularly effective to have students participate directly in the drawing of random samples from a population and the construction of an empirical sampling distribution. For example, you can bring in a population of pennies, have students randomly select five pennies from this population, and then determine the average age of these five pennies. Next, you can pool these averages across the class and construct a dot plot or histogram of the distribution (see Scheaffer et al. 2004). Or, you can bring in a large population of M&M's or similar candies, have each student draw a different sample of twenty-five candies, and determine the proportion of one color. Then, you can pool these sample proportions across the class and construct a dot plot of the sample proportions (see Rossman and Chance 2001). Along the way, you can have students construct and examine graphs of all *three* distributions: the population, their samples, and the empirical sampling distribution for the entire class. For example, in the candy-sampling activity, you might get graphs such as those in figure 7.6.

Discussion should focus on the pattern of each distribution and how these distributions compare for the population, the individual samples, and the sampling distribution. There is a familiar pattern in that the sampling distribution displays less variability than the population and the individual samples. Once students feel comfortable with the definition of a sampling distribution, you can use technology to draw many more samples and to explore the effect of sample size, population shape, population size, and so on, as illustrated in Example 3, an activity that can be used directly with your students.

EXAMPLE 3: AGES OF BILLIONAIRES

This example indicates the type of progression through which you might want to lead your students, ideally having them do the exploration and creation of graphs themselves. The Data and Stories Library (DASL) at lib.stat.cmu.edu/DASL/Datafiles/Billionaires92.html has data on 233 bil-

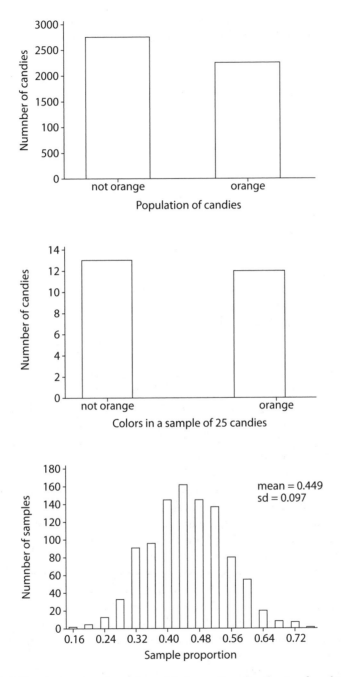

Fig. 7.6. Population, sample, and empirical sampling distribution for a binary variable

lionaires (or billionaire families), including their wealth and age in 1992. Figure 7.7 shows the distribution of the ages of the billionaires.

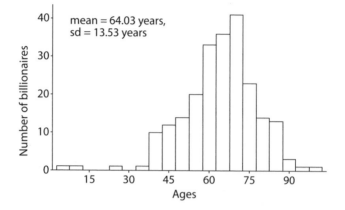

Fig. 7.7. Frequency distribution for ages of billionaires

This distribution of ages is fairly symmetric, with some low outliers (child actors?). We would expect a sample from this population to have similar characteristics.

Figures 7.8, 7.9, and 7.10 show empirical sampling distributions of the sample mean ages for 500 samples of billionaires drawn from this population for sample sizes $n = 1$, $n = 5$, and $n = 20$, respectively. In these graphs, each observation is the mean of one sample of that sample size.

Notice that figure 7.8 looks a lot like the original population. Since $n = 1$, it is the distribution of individual observations, and it behaves like one very large sample. Thus, when $n = 1$, the sampling distribution of the sample means will have similar characteristics to the population. If we took more samples, the mean and standard deviation would get even closer to the population values. In figure 7.9, the distribution of the sample means is more symmetric, but the center is still around 64 years. The variability in this distribution is much smaller, as the sample means cluster more closely around the population mean than the individual observations did in figure 7.8.

The variability of the sample means decreases further when the sample size is raised to $n = 20$. Now the distribution is becoming very symmetric, with its mean again around 64 but with a much smaller standard deviation. Keep in mind that each element being tabulated in figure 7.10 represents the average of 20 billionaire ages. If we were to examine just the distribution of one of these 500 samples, we would see that it has a slight skew to the left, like the population, with a mean and standard deviation similar to

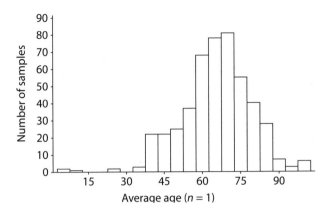

Fig. 7.8. Mean ages for 500 samples of billionaires of sample size 1

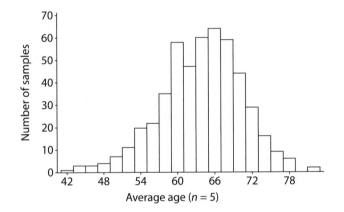

Fig. 7.9. Mean ages for 500 samples of billionaires of sample size 5

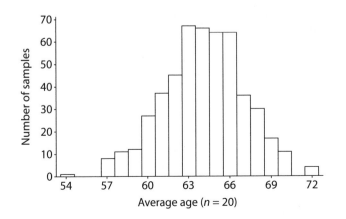

Fig. 7.10. Mean ages for 500 samples of billionaires of sample size 20

that of the population. Fig. 7.11 displays the frequency of the data on the total wealth of the same billionaires.

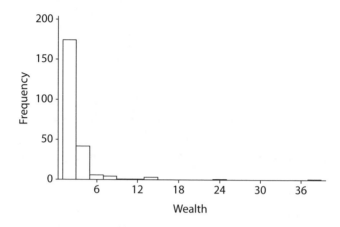

Fig. 7.11. Frequency distribution of total wealth in billions

This distribution of total wealth is strongly skewed to the right with mean = 2.68 (billion dollars) and standard deviation 3.32 (billion dollars). Figure 7.12 shows three empirical sampling distributions of the means for the wealth of 500 samples of sample sizes 1, 5, and 20 (figs. 7.12a, 7.12b, and 7.12c, respectively).

We see a similar progression in these data to what we saw in figures 7.8, 7.9, and 7.10 with the average age data. When $n = 1$, the sampling distribution looks like the population. As n increases, the distribution of the means becomes less skewed and the standard deviation decreases. In this instance, with $n = 20$, the distribution is not yet symmetric. Figure 7.13 shows the empirical sampling distribution for samples of size $n = 50$.

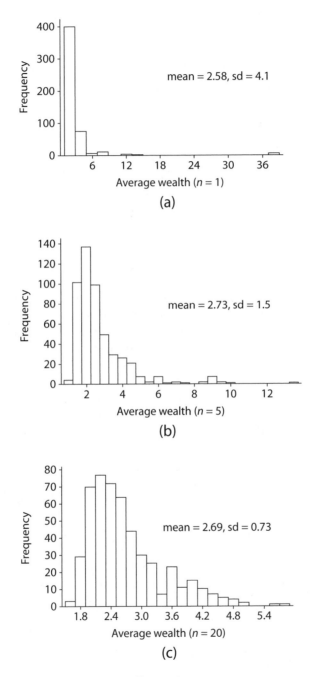

Fig. 7.12

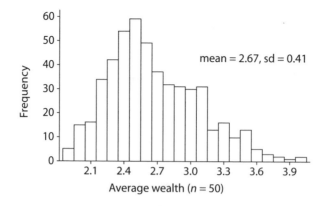

Fig. 7.13. Sampling distribution of total wealth of billionaires for samples of size
n = 50

Although there is still a slight skew to the right, this distribution is much more symmetrical than those shown in figure 7.12. With such a large sample size, the sample means will all fall very close to the true population mean. Thus, even when the population distribution is heavily skewed, the sampling distribution of sample means will eventually look symmetric *if* the sample size is large enough. How large *n* has to be depends on the severity of the skewness in the original population. After seeing several examples, like Examples 2 and 3 above, students should begin to realize that it doesn't matter what the distribution of the original population distribution looks like, since the sampling distribution of the sample means will eventually look symmetrical as the sample size *n* becomes larger.

In figure 7.13, if we looked at the distribution of only one of the samples of 50 billionaires, that distribution would be strongly skewed to the right, with its mean and standard deviation similar to those of the population.

SUMMARY: POINTS TO KEEP IN MIND

- When *n* = 1, the sampling distribution will resemble the population. A consequence of this is if the population is normal to begin with, the sampling distribution will always look normal. If the population is not normal, then even samples of size 2 will lead to a sampling distribution with a slightly different shape than the population distribution. For example, if we consider a population with just two possible outcomes, the theoretical sampling distribution (means from *all* possible samples) will have three outcomes: the original two values and their average.

- When the population is much larger than the sample, the actual size of the population does not affect the sampling distribution. As an analogy, if we want to decide if we like the taste of a new soup, as long as the soup's ingredients are well mixed, if doesn't matter if our spoonful is coming from a dinner bowl or from the twenty-gallon kettle! One spoonful will give us a pretty good idea of whether or not we like the soup.

- When we have a large enough sample size for the sampling distribution to be symmetric, we can use the empirical rule to predict where those sample statistics should lie. In particular, 95 percent of observations (in this situation, sample statistics) should fall within 2 standard deviations of the sampling distribution mean. It is important for students to practice applying this empirical rule to sample statistics instead of just to individual data values.

- Students often confuse, or get sloppy with distinguishing between, the "number of samples" and the "sample size." Increasing the sample size results in a sampling distribution that is less spread out and more normal-shaped, but increasing the number of samples just means that the empirical sampling distribution will better match the theoretical sampling distribution.

- Students need to realize *why* we look at sampling distributions. In real life, we usually just take one sample. However, in order to decide whether a sample statistic is unusual or typical, we need to understand the long-term pattern of the values of that sample statistic. Such information is provided by sampling distributions.

- Sampling distributions are at the heart of how we make predictions in statistics. In particular, sampling distributions relate to how we determine the margin of error and how we calculate *p*-values. For further discussion, see chapter 8 ("What Is a Margin of Error?") and chapter 9 ("What Is a *p*-Value?").

8

What Is a Margin of Error?

To begin this discussion, we shall continue the discussion of the GPA question, which was introduced in the previous chapter.

EXAMPLE 1: GRADE POINT AVERAGES

The population consists of GPA values with mean 3.23 and standard deviation 0.276. The graph in figure 8.1 displays our empirical sampling distribution based on 500 samples of size 25. The normal distribution predicted by the central limit theorem is superimposed, with mean 3.23 and standard deviation $0.276/\sqrt{25} = 0.055$.

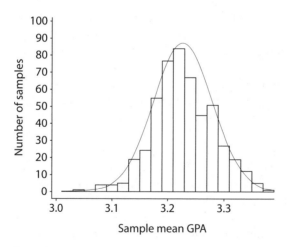

Fig. 8.1. Sampling distribution for GPA, sample size 25

The *model* of the normal distribution does a fairly good job of describing the behavior of these sample means. Using the normal distribution, we can apply the empirical rule: 95 percent of the observations should fall within 2 standard deviations of the mean. Here the observations are sample means, the mean of the distribution of sample means is 3.23, and the standard deviation is 0.055. So approximately 95 percent of sample means should fall inside the interval $(3.23 - 2(0.055), 3.23 + 2(0.055)) = (3.12, 3.34)$, as depicted in figure 8.2.

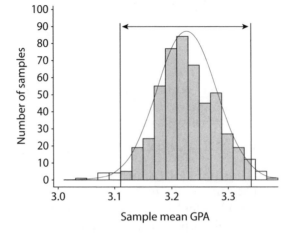

Fig. 8.2. A 95% confidence interval for GPA

For our actual simulated sample means, 480 of the 500 samples, or 96%, are within this interval and within 2 standard deviations $(2 \times 0.055 = 0.11)$ of the population mean. This value, 0.11, is called the "margin of error." The margin of error indicates how far we can expect our sample results to be from the population result, taking into account the inevitable variability that occurs from random sample to random sample. For the 500 samples selected above, if we add and subtract 0.11 from each sample mean, we obtain 500 intervals. For this simulation, 472 of the 500 intervals, represented by the gray intervals in figure 8.3, contain the value 3.23 between the two endpoints. We can see that although the intervals change from sample to sample, all but the 28 black intervals contain the population mean (we'd expect 95% of them would).

Now think about the situation where we don't know the population mean, but the population standard deviation is still about 0.276. If we take 500 random samples of twenty-five students and find the sample mean for each student, we still expect roughly 95 percent of the sample means to be within 0.11 point of the population mean GPA (whatever value that is). Thus, when we take just one sample (which is the situation in reality), we have some faith that that sample mean will be within 0.11 point of the population mean. Since this statement is accurate 95 percent of the time, the best we can say is that we are *95* percent *confident* that the population mean is within 0.11 point of the sample mean we obtain.

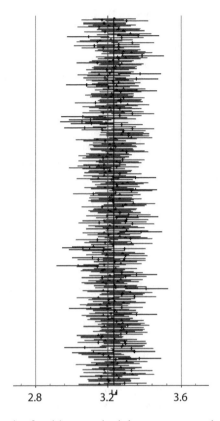

2.8   3.2   3.6

Fig. 8.3. Intervals of width 2 standard deviations around sample means

CATEGORICAL DATA

The same logic can be applied when working with a sample proportion rather than a sample mean. The central limit theorem predicts that the sampling distribution of sample proportions will be well modeled by a normal distribution with its mean equal to the proportion of successes in the population, p, and its standard deviation equal to $\sqrt{p(1-p)/n}$. As above, if we take a large number of random samples, we would expect 95 percent of the sample proportions to fall within a margin of error of 2 standard deviations, or $2\sqrt{p(1-p)/n}$, of the population proportion. The function $f(p) = 2\sqrt{p(1-p)/n}$ has derivative $f'(p) = (1-2p)/\sqrt{p(1-p)/n}$, which has a critical point, a maximum, at $p = .5$. If we substitute $p = .5$ into the margin-of-error formula, we obtain $f(.5) = 1/\sqrt{n}$.

Thus the maximum value for the margin of error is $1/\sqrt{n}$ when we estimate a population proportion.

EXAMPLE 2: POLLING

The term *margin of error* arises most commonly in polls or surveys. In October 2003, the Gallup agency conducted a survey of 1017 adult Americans and asked whether they would describe the problem of drugs as extremely serious, very serious, moderately serious, not too serious, or not serious at all. The agency found that 71 percent rated the problem as extremely serious or very serious. If the preceding argument is used, the margin of error of this poll can be approximated by the formula $1/\sqrt{n}$ = .03. Thus, if we were to take many random samples of 1017 adult Americans, we would expect 95 percent of the sample proportions to fall within .03 of the true population proportion. The conclusion from the Gallup survey is that we are 95 percent confident that the true proportion of adult Americans who consider drugs to be a very or extremely serious problem is between .71 − .03 and .71 + .03, or in the interval (.68, .74).

SUMMARY: POINTS TO KEEP IN MIND

- Margin of error measures the "error" due to variability in random sampling. It is really not an "error" at all; rather, it stems from sampling variability. Margin of error does not measure any errors caused by a poorly collected sample, poor wording of the question, dishonest answers, and so on. It simply provides an interval within which the population value plausibly lies. (A better term might actually be "margin of success"! But statisticians probably would protest that term!)

- The $1/\sqrt{n}$ formula gives us a quick approximation of the margin-of-error window for a sample proportion. More formally, the margin of error of a sample proportion is 2 standard deviations, $2\sqrt{p(1-p)/n}$, where p equals the population proportion. Since we don't know the true value of p, we often estimate the value of this margin of error by using the sample proportion as an estimate for the population proportion p. Taking larger samples reduces the margin of error because the standard deviations will be smaller. This allows us to make a more *precise* estimate of the population value.

- In working with sample means, we used $\sigma/\sqrt{n}$ above. Typically we don't know σ, population standard deviation, and so instead we use $s/\sqrt{n}$, where s is the sample standard deviation, as an estimate. This change necessitates a slightly more complicated margin-of-error calculation, but the logic is the same.

This margin-of-error calculation is valid only if the conditions necessary for the central limit theorem are met. These conditions include (1) a random sample from the population of interest, (2) a constant probability of success and a large sample size for categorical data, and (3) either a large sample size or a normally distributed population in the case of quantitative data.

9

What Is a *p*-Value?

A *p*-value is a probability, and it measures the strength of evidence against some hypothesis. The smaller the *p*-value, the stronger the evidence is against the hypothesis. By convention, a *p*-value below .05 reflects evidence strong enough to reject the corresponding hypothesis. In statistics, we typically take a sample from a large population, and we want to make some decision about the population, based on what we observe in the sample and the corresponding *p*-value of these data.

Example 1: An Empirical Example, Using Simulation

Suppose a student selects a handful of 10 candies from a well-mixed jar of 1000 candies. The student has been told the jar contains 500 red and 500 blue candies. She is very surprised that in her 10 candies, she found only 2 red candies and begins to question whether the information she was given of an equal split in the population can be accurate. From her sample of 10 candies, can she refute the hypothesis that there are 500 red and 500 blue candies in the mixture?

In order to answer this question, we want to know the behavior of random samples from the specified population. If the population truly has 500 red candies, is it surprising for a random sample to yield only 2 red candies? One way to answer this question is to use a computer to simulate drawing many random samples from such a population and to keep track of how many red candies we obtain each time.

The chart in figure 9.1 shows the results of a simulation of 50 samples of size 10 drawn from the hypothesized 500 red population, using the software package ProbSim (Konold and Miller 1990).

We see that in a population of 500 red and 500 blue candies, the most common outcomes for a sample of 10 candies were 4, 5, or 6 red candies and that very few samples contain at least 8 or at most 2 red candies. According to these results, a sample of at most 2 red candies happens approximately 4 percent of the time when we randomly select 10 candies from a population that is 50 percent red. This outcome provides some evidence that either the jar was not well mixed when the student took her sample or that her sample actually came from a population with fewer red candies to begin with. For example, suppose we thought the population had 300 red candies and 700 blue candies. Figure 9.2 shows the results of drawing 50 random samples from a 300 red and 700 blue population.

Number of Red Candies	Frequency	Relative Frequency	
10 red, RRRRRRRRRR	0	0	
9 red, e.g., RRRRRRRRRB	1	.02	*
8 red, e.g., RRRRRRRRBB	2	.04	**
7 red, e.g., RRRRRRRBBB	8	.16	********
6 red, e.g., RRRRRRBBBB	12	.24	************
5 red, e.g., RRRRRBBBBB	11	.22	***********
4 red, e.g., RRRRBBBBBB	9	.18	*********
3 red, e.g., RRRBBBBBBB	5	.10	*****
2 red, e.g., RRBBBBBBBB	1	.02	*
1 red, e.g., RBBBBBBBBB	1	.02	*
0 red, BBBBBBBBBB	0	0	

Fig. 9.1. A sampling distribution for the number of red candies in samples of size 10 in a 50% red mixture

Number of Red Candies	Frequency	Relative Frequency	
10 red, RRRRRRRRRR	0	0	
9 red, e.g., RRRRRRRRRB	0	0	
8 red, e.g., RRRRRRRRBB	0	0	
7 red, e.g., RRRRRRRBBB	0	0	
6 red, e.g., RRRRRRBBBB	1	.02	*
5 red, e.g., RRRRRBBBBB	7	.14	*******
4 red, e.g., RRRRBBBBBB	16	.32	****************
3 red, e.g., RRRBBBBBBB	8	.16	********
2 red, e.g., RRBBBBBBBB	11	.22	**********
1 red, e.g., RBBBBBBBBB	5	.10	*****
0 red, BBBBBBBBBB	2	.04	**

Fig. 9.2. A sampling distribution for the number of red candies in samples of size 10 in a 30% red mixture

If the population was actually 30 percent red, then no one would have raised any eyebrows at drawing a sample of 10 candies with 2 red. Approximately 18 percent of random samples have 2 or fewer red candies. That is, the one-sided *p*-value is approximately .18. Since her outcome is not surprising, her sample does not provide evidence against a claim of 30 percent red.

In summary, under the hypothesis "50% red," the probability of getting a sample like the one the student obtained is rather small. Since the teacher promises the jar was well mixed, the student has some evidence to reject the claim that the jar's composition was truly 50% red. Of course, it's also possible that the student was just unlucky. But a population proportion of .30 seems much more plausible.

The logic that justifies the rejection of a 50% red mixture is an extension of a familiar form of reasoning; for example, "If it is warm outside, then the people I see when I look out the window will not be wearing coats. In fact, the people I see *are* wearing coats. Therefore, I have strong evidence that it is not warm outside." Two things can make the statistical form of this logic hard to grasp: (1) reasoning from probability distributions, and (2) technical baggage. The probability can't be avoided, but the technical baggage can be kept to a minimum, especially at first. That baggage includes a lot of specialized vocabulary (*null* and *alternative*, *level* and *power*, *Type I* and *Type II*, *one-sided* versus *two-sided*) and a variety of formulas for computing *p*-values. However, the fundamental logic is not much harder than looking out the window to see if you need to put on your jacket.

A *p*-value answers a question about the relationship between data you observe and some hypothesis about the data: "If the hypothesis is true, how likely would it be to get data like the data I actually got?" Less formally, "If I believe the hypothesis, how surprised should I be to get data like these?"

Hypothesis: It is warm out.

Data: People are wearing coats.

p-value: Suppose it is warm out. Then the chance that people will be wearing coats (the *p*-value) is small.

Conclusion: Reject the hypothesis.

Sherlock Holmes uses this logic in the story of Silver Blaze, a racehorse who is stolen just before a big race. Holmes points out the "curious incident of the dog in the night-time": the dog had not barked at the time the horse was stolen. Here is Holmes's reasoning, recast in the language of hypothesis testing. "My hypothesis is that the horse was stolen by a stranger. My observed data indicate 'no bark.' If my hypothesis is true, and a stranger was in the stable, then 'no bark' is highly unlikely, and my *p*-value is near 0. So I reject my hypothesis and conclude that it was not a stranger who took the horse."

COMPUTING *P*-VALUES

To compute a *p*-value, you must first have two things: (1) observed data values, and (2) a specific hypothesis (called the *null hypothesis*) that tells which data values are likely and which are not under that hypothesis. The *p*-value is the probability assigned to the actual observed data by your hypothesis. A bare-bones example follows.

EXAMPLE 2: COIN FLIPS—COMPUTING A P-VALUE

Scenario: To test a coin for fairness, you flip it 10 times, and get heads every time.

Null hypothesis: The coin is fair; flips are independent.

Observed data: In 10 flips, all landed heads.

p-value: $P(10 \text{ heads in 10 flips of a fair coin}) = P(H) \times P(H) \times \ldots \times P(H) = (1/2)^{10} \approx .001$.

With such a small *p*-value, we have strong evidence that this is not a fair coin when flipped.

INTERPRETING *P*-VALUES

The next example is a class demonstration that can help put students in touch with the basic intuitive interpretation of *p*-values. This example also foreshadows the formal use of *p*-values in statistical inference.

EXAMPLE 3: CALLING COIN FLIPS—*P*-VALUES MEASURE SUSPICION

Scenario: Show your class a fair coin. Prepare to toss it, but first, ask a student to call the outcome, heads or tails. Then flip the coin, look at the result, and announce, "Right," regardless of the outcome. Ask the same student to call another one, then flip, and again announce, "Right." Continue with calls and flips, always announcing, "Right," regardless of the actual outcome. At the point when students begin to get restless, ask them to discuss when they first began to get suspicious, and why. Students should realize intuitively that the observed outcomes cast suspicion on the implicit null hypothesis that the tosses are fair, the teacher honest, and the caller just lucky.

Analysis: Here the null hypothesis is that, because the coin is fair and flips are independent, the student has a 50-50 chance of calling each flip correctly. (There are several alternative explanations, e.g., the teacher is able to manipulate the outcome of the toss, the teacher is lying, the student is clairvoyant, etc., but what matters for the *p*-value is that "if nothing is going on," the chance of a correct call is 50-50.) For a sequence of five tosses, the probability of getting all five right is $(1/2)^5$ ª .03. [*Null hypothesis*: P(Correct call) = 1/2. *Data*: Five in a row correct. *p*-value: $(1/2)^5$. For a sequence of ten correct calls in a row, the *p*-value is $(1/2)^{10} \approx .001$.]

The logic here is "either the null hypothesis is false or something extremely unlikely has occurred." In Example 3, students' skepticism builds as the string of "correct" guesses lengthens and the *p*-value shrinks toward zero: the stronger the evidence against the null hypothesis, the smaller the *p*-value. Several complications can make it harder for students to achieve and retain a clearer sense of the basic idea. The examples that follow illustrate four of these complications.

p-Values Are Tail Probabilities

In Example 2, the observed data (10 heads in all 10 flips) correspond to the strongest possible evidence against the null hypothesis (that the coin is fair when flipped). It often happens that the outcome we observe is not as extreme as other outcomes we might have gotten. The *p*-value measures the probability of an *outcome at least as extreme as the one we actually got*. (Probabilities of this form are called *tail probabilities*.)

EXAMPLE 4: EYE DOMINANCE I—AT LEAST AS EXTREME

Scenario: Just as most people are either right-handed or left-handed, for most people one eye tends to be dominant. (To check your own eyes, form a two-inch circle with your hands held at arms length, and look through the circle at an object on the far side of the room. Then close one eye. If the object you picked is no longer in the circle, the eye you closed is your dominant eye.) Now suppose you want to check whether the right eye tends to be dominant among right-handed people. You test 15 right-handers for eye dominance, and find that 11 are right-eye dominant.

Null hypothesis: Subjects are independent; among right-handers, right and left eyes are equally likely to be dominant.

Data: 11 of 15 are right-eye dominant.

p-value: P(11 or more of 15 are right-eye dominant *if* dominance is
   equally likely to be right eye or left eye)
   = P(11 *or more* heads in 15 tosses of a fair coin)
   = $P(11) + P(12) + \ldots + P(15)$
   $\approx .06$, using the binomial distribution.

Thus, the probability of getting at least 11 right-eye dominant people in a sample of 15 right-handers, if the right-handers were equally likely to be right-eye dominant and left-eye dominant, is .06. This provides moderate evidence against the null hypothesis, in favor of right-eye dominance.

One or Two Tails?

In Example 4, "at least as extreme" means "11 or more." Often, "at least as extreme" is two-sided, because both unusually large and unusually small values cast suspicion on the null hypothesis.

EXAMPLE 5: EYE DOMINANCE II—TWO-SIDED P-VALUES

Scenario: In the population as a whole, are left and right eyes equally likely to be dominant? Suppose you have 15 randomly chosen subjects, of whom 11 are right-eye dominant.

Null hypothesis: Subjects are independent; right and left are equally likely.

Data: Out of 15, 11 are right, and 4 are left.

p-value: $P(11$ or more, or 4 or fewer *if* no eye dominance)
$$= [P(0) + \ldots + P(4)] + [P(11) + \ldots + P(15)]$$
$$\approx .12.$$

If people were equally likely to be right-eye and left-eye dominant, how often would we see at least this many or at most this few right-eye dominant people in the sample? This is not a super surprising outcome (*p*-value > .05), and we would say we don't have convincing evidence against the hypothesis that left and right eyes are equally likely to be dominant.

MULTIPLE DATA VALUES: WHAT DOES *EXTREME* MEAN?

In the examples so far, there has been only a single observed value for the data in question, that is, the data have been univariate. This makes it easy to see what *extreme* means. When we have multiple values that can't be automatically summarized using a single number, we have to specify (define/invent/choose) what we mean by *extreme*. In many courses, this first becomes an explicit issue when students encounter the chi-square test for contingency tables.

EXAMPLE 6: WEALTH AND POLITICAL POWER—MEASURING "EXTREME"

Scenario: The table in figure 9.3 classifies residents of seventeenth-century Salem Village, Massachusetts, according to wealth, as measured by their tax assessment for a year and whether or not they were ever elected to the Village Committee over a period of time. Were wealthier people more likely to be elected?

Tax (shillings)	Terms on the Village Committee		Total	% ≥ 1
	0	≥ 1		
0 to 120	16	0	16	0.00%
120+ to 240	9	2	11	18.18%
240+ to 480	24	9	33	27.27%
480+ to 960	9	12	21	57.14%
960+ and up	0	11	11	100.00%
Total	58	34	92	36.96%

Fig. 9.3. Wealth and power in Salem Village, Massachusetts

Null hypothesis: There is no evidence of association between wealth and power—the data look like the sort you might get by putting names and tax assessments on slips of paper and "electing" 34 of them by drawing at random from a hat.

p-value: The *p*-value is the probability that if you draw at random, you would get data at least as extreme as the actual data in the table.

Comment: Here, it isn't obvious what "at least as extreme" means. The standard approach is this: First, compute a set of "expected values"—what you would get if the percentage elected in each row exactly matches the overall percentage, here 34/92 ⇒ 37%. Then compute, for each cell of the table, (observed − expected)^2/expected, and add up these values across all the cells in the table. The resulting value, called the chi-square statistic, is a measure of how far the data values are from the values you would expect if the null hypothesis were exactly true. The *p*-value is the probability of getting a chi-square value this large or larger.

In figure 9.3, the last column shows that the percentage elected is larger for the categories of larger tax assessment. The p-value turns out to be less than .001, presenting very strong evidence against the null hypothesis. We conclude that there is evidence of an association between wealth and power for this population. The data in the figure are quite unlikely to occur by random chance alone.

REMINDERS AND PITFALLS

The different complications make it harder for a teacher to help students keep their focus on the logic that connects the null hypothesis and the p-value that follows from it. Here are two useful reminders. (1) The null hypothesis must convey information about the type of data we expect to see. (For example, the hypothesis of warm weather tells us that it is unlikely we will see lots of people wearing coats.) (2) The p-value gives the probability of the data (what you get to see) and *not* the probability of the hypothesis being true (although this is really what you'd like to know). As long as either of these statements seems obscure or confusing, students haven't yet assimilated the logic of p-values. In particular, they may fall victim to a common fallacy, sometimes called the prosecutor's fallacy.

The Prosecutor's Fallacy: "The p-value is .001. This means that the chance is only 1 in 1000 that the null hypothesis is true."

In classical statistics, probabilities are assigned only to data; probabilities are never assigned to hypotheses or values of unknown parameters. The reasoning is that the data are what is random, changing as we take different samples, whereas population parameters are fixed, though we might not know the exact values for them. The fallacy comes from interpreting the conditional probability $P(\text{data} \mid \text{hypothesis})$ as the conditional probability $P(\text{hypothesis} \mid \text{data})$. Because the null hypothesis tells which data values are likely and which are not, it should make sense that we can compute probabilities in the direction "hypothesis $\rightarrow$ data," that is, $P(\text{data} \mid \text{hypothesis})$. At least implicitly, the null hypothesis "contains instructions" for computing the chances for the different possible data outcomes. The data values contain no instructions for computing any sort of probabilities, and so, in particular, it is impossible to compute probabilities in the direction "data $\rightarrow$ hypothesis." The data alone can't tell us how to compute the probability that a particular hypothesis is true. As the Berkeley statistician David Freedman and his coauthors might have put it in their classic book, *Statistics*, (Freedman, Pisani, and Purves 1978), the chance is in the data, not the hypothesis.

One way to help students remember this idea is to connect it to the relative frequency definition of probability, as we did in our initial example in this chapter. We make probability statements about how often events occur if we repeat the same process over and over. Here, we consider the random selection of the sample or the random assignment of individuals to treatments as repeatable. The null hypothesis indicates how often we expect to see different types of results. So what changes each time are the observed results, not the null hypothesis. We can assign probabilities to results but not to hypotheses.

Summary

It is useful to link the logic of a *p*-value to one's intuition about likely and unlikely scenarios under certain conditions. Students can be led through a scenario where they will tell us that "this wouldn't happen by chance if…." It is important to reinforce that intuition before continuing on into the formal probability calculations and technical details. As students' comfort with these ideas increases, it will always be important to focus on what the *p*-value says and what it does not say.

10

What Are Degrees of Freedom and How Do We Find Them?

Consider the following analogy from algebra that mirrors degrees of freedom in statistics. Suppose we are given a linear equation in three variables of the form $3x + 4y - 2z = 100$. If the values of any two of the variables in this equation are known, the third value is automatically also known. For example, if $x = 4$ and $y = 2$, then z would have to be -40. There isn't a "choice" for z, if x and y are known. Any two of the variables in the equation $3x + 4y - 2z = 100$ are "free to roam," but the third is determined by the other two. In algebraic language, one would say that there are "two linearly independent variables" in the equation, the third variable being dependent on the first two. In statistical language, there are "two degrees of freedom" in this linear equation in three variables. This example draws a helpful analogy to how degrees of freedom arise in statistics.

THE CASE IN STATISTICS

A similar phenomenon occurs in statistics. Suppose that a company is starting a new branch in another city and will be hiring 100 entry-level people, 50 mid-level managers, and 40 top-level executives to run the new operation. Both men and women are hired at all three levels, and the company has promised to hire an equal number of men and women. When the hiring process is completed, the question arises whether gender and level are independent in the hiring process. The question suggests the following analysis (see the table shown in fig. 10.1).

Level/Gender	Men	Women	Total
Top Level			40
Mid Level			50
Entry Level			100
Total	95	95	190

Fig. 10.1. Job-level data in hiring a group of men and women

The existing constraints from figure 10.1 are the total number of positions available at each level (100, 50, 40), and an equal split (95 and 95) in the number of hires of men and women. How many entries could be made in the table before all the other entries would necessarily be completely determined?

As seen in the table in figure 10.2, if 27 men were hired at the mid-level, **23** women would have also been hired at that level, but the frequencies in the other four slots are as yet undetermined.

Level/Gender	Men	Women	Total
Top Level			40
Mid Level	27	**23**	50
Entry Level			100
Total	95	95	190

Fig. 10.2. Job-level data in hiring a group of men and women with a known value

Given even further information, let us say that 47 women were hired at entry-level positions as shown in figure 10.3, then the rest of the entries (*) are completely determined by the known constraints in the margin totals and *must* then be as shown in figure 10.4. The entries in bold in these tables were completely determined by two pieces of information—that there were 27 mid-level men and 47 entry-level women. There is nothing special about the two entries that were known, any two of the six entries in the table will completely determine the other entries. Thus, we have only "two degrees of freedom" in this table—two "free" values, so to speak.

Level/Gender	Men	Women	Total
Top Level	*	*	40
Mid Level	27	**23**	50
Entry Level	*	47	100
Total	95	95	190

Fig. 10.3. Job-level data in the hiring of a group of men and women, with several known values

Level/Gender	Men	Women	Total
Top Level	15	25	40
Mid Level	27	23	50
Entry Level	53	47	100
Total	95	95	190

Fig. 10.4. Complete job-level data in hiring a group of men and women

DEGREES OF FREEDOM IN TWO-WAY TABLES

A typical statistical test for this type of situation might be to calculate a chi-square statistic with two degrees of freedom, to test for independence of gender and job level.

To examine the *relationship* between two variables from a two-way table, we calculate the chi-square statistic:

$$\chi^2 = \Sigma_{ij} \frac{\left(\text{observed}_{ij} - \text{expected}_{ij}\right)^2}{\text{expected}_{ij}}$$

In our hiring example from above, there are the same restrictions to the values in this sum, determined by the fixed row and column totals. Thus, the degrees of freedom of this sum are the same as for the table in figure 10.1. The sampling distribution of this statistic is approximated by a chi-square distribution with two degrees of freedom (see fig. 10.5). This tells us that a χ^2 value larger than 5.99 will be judged statistically significant at the .05 level. If we had a larger two-way table, several more terms would be added together to calculate this statistic, and then the χ^2 value would need to be even larger before we would find it "surprising" to happen by chance alone. For example, with nine degrees of freedom, the χ^2 value would have to exceed 16.92 to be significant at the .05 level. Thus, the degrees of freedom determine which distribution we use to model the sampling distribution of the sample chi-square statistic.

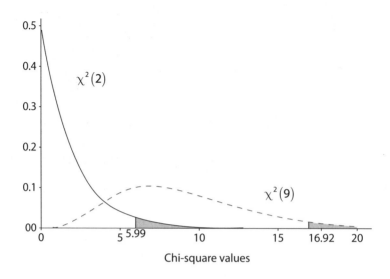

Fig. 10.5: Chi-square distributions with two and nine degrees of freedom, respectively

As another example of degrees of freedom, consider a situation in which a data set has a mean of $\bar{x}$ for n data points, $x_1, x_2, x_3, \ldots, x_n$. If we already know the value of the mean, we would need to know only the values of $n - 1$ of the data points in order to determine the value of the nth data point. This is a generalization of our first example to $n - 1$ linearly independent entries. Similarly, to calculate the standard deviation of a sample, we must first calculate the sample mean, and this imposes a restriction on the values in the sample, since the sum of the values divided by n must equal the sample mean. Thus there are $n - 1$ degrees of freedom in this instance as well, arising from the fact that we used the sample mean $\bar{x}$ in the standard deviation formula as an estimator for the (unknown) population μ. In fact, any time we have to estimate a parameter from the data set, we lose a degree of freedom. In simple linear regression, we estimate the slope and the intercept, so the associated degrees of freedom are $n - 2$ for n observations in a linear regression. If we actually had data for an entire population (rather than just a sample) and if we knew the population μ, we would have n degrees of freedom when calculating the standard deviation of the population, since we wouldn't lose a degree of freedom as we do when calculating a sample mean. (See chapter 11 on the question "Why divide by $n - 1$?" for some further discussion.)

DEGREES OF FREEDOM IN THE t STATISTIC

In estimating or making inferences about the sample mean, whenever we use the sample standard deviation as an estimate for the population standard deviation, we calculate a t statistic as $(\overline{x} - \mu)/(s / \sqrt{n})$, which also has $n - 1$ degrees of freedom, since the nth data value is determined once we know $n - 1$ values and the mean. Note that if we had known the population standard deviation, then the test statistic $(\overline{x} - \mu)/(\sigma / \sqrt{n})$ would not have any limitations on the data values (and follows a standard normal distribution). When we substitute in s for σ, we lose one degree of freedom and have an additional source of sampling variability ($\overline{x}$ and s). For these reasons, the t statistic is well modeled by the Student t distribution, which takes the degrees of freedom into account, as did the chi-square distribution above, and has heavier tails than the standard normal distribution. But as the degrees of freedom increase (reflecting a larger sample size and therefore a more precise estimate of σ), the t distribution loses the heaviness in the tails and approaches the normal distribution (see fig. 10.6). A t distribution with infinite degrees of freedom is equivalent to the standard normal distribution. Consequently, with larger sample sizes, a t statistic value does not need to be as extreme to be considered surprising.

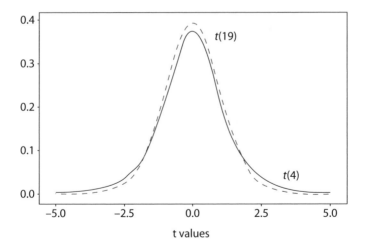

Fig. 10.6. t distributions with four (solid line) and nineteen (dashed line) degrees of freedom

CONCLUSION

In summary, degrees of freedom arise in statistics when one is working with sample data, because of constraints placed on the observations in the sample. Knowing the degrees of freedom supplies us with information about the amount of sampling variability. Inferential techniques must take degrees of freedom into account when we are judging the extremeness of the sample results.

11

Why Do We Divide by $n - 1$ instead of n?

In teaching students about standard deviation, we'd like for them to develop an intuitive notion of "average deviation from the mean." As we've seen briefly in chapter 3, "Why Are Deviations Squared?" we look at how far each observation is from the mean, square those deviations, and then "average" them. So, if we have a population of N observations with mean μ, we can define the *variance* to be

$$\sigma^2 = \frac{\sum_{i=1}^{n}(x_i - \mu)^2}{N}.$$

Intuitively, this is an average squared deviation from the mean for the population; thus, it is the *population variance*. However, when we take a sample of size n from this larger population, the most common definition of the variance for the *sample* is

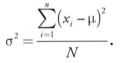

$$s^2 = \frac{\sum_{i=1}^{n}(x_i - \bar{x})^2}{n - 1}.$$

Why?

THE BEHAVIOR OF ESTIMATORS

One way to explore this issue is to remember that the goal of taking a random sample is to make inferences about the population. Suppose we have a population with variance σ^2. Is s^2 a reasonable estimate of σ^2? What does it mean to be a reasonable estimate? It means that the sample results will tend to be close to the population value. There are two things to consider when measuring "close" here: (1) Are the sample values clustered around the population value we are trying to estimate? (2) How large is the variability in the sample values from sample to sample?

We can explore these questions through simulation. Let's take lots of samples from a population where we know a value for σ^2, calculate s^2 for each sample, and see if they tend to be close to σ^2. That is, let's explore the sampling distribution of s^2. Figure 11.1 shows a sampling distribution for 5000 values of sample variances, where the samples were pulled from a population with known population mean and known population variance.

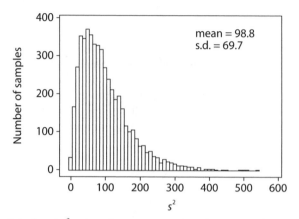

Fig. 11.1. Calculated s^2 values for 5000 samples of size $n = 5$ from a normal distribution with $\sigma^2 = 100$

Thus, s^2 seems to be a pretty good estimator of σ^2—the values of s^2 cluster around the true value of 100 (the empirical mean of the sample variances is close to the population variance). As we expect, if we increase the sample size to $n = 10$, the estimates cluster even more closely around σ^2, as can be seen in figure 11.2.

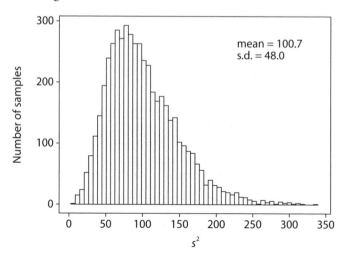

Fig. 11.2. Calculated s^2 values for 5000 samples of size $n = 10$ from a normal distribution with $\sigma^2 = 100$

What if we divided by n instead of $n-1$ in the formula for sample variance? How well would those values cluster around the population variance? Would they be better or worse than when we divided by $n-1$, as above? That is, what if we used the formula

$$s_n^2 = \frac{\displaystyle\sum_{i=1}^{n}(x_i - \bar{x})^2}{n}$$

as an estimator for σ^2 instead? Figure 11.3 shows the results for samples of size 5 and 10.

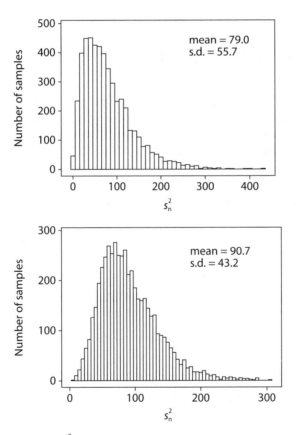

Fig. 11.3. Calculated s_n^2 values for 500 samples from a population with $\sigma^2 =$ 100 with samples of size $n = 5$ and with samples of size $n = 10$

Again, we see that the estimates cluster closer together when we increase the sample size (less variability), but they are not exactly clustering around $\sigma^2 = 100$ as they did when we divided by $n - 1$. They appear to have a tendency to underestimate the value of σ^2. In fact, we could show analytically that the expected value of the estimator obtained from dividing by n is equal to $E\left(S_n^2\right) = \left((n-1)/n\right)\sigma^2$. Thus, in our example above, instead of getting the population mean 100 as the expected value of our estimator, we would get $(4/5)(100) = 80$ for the samples of size 5 and $(9/10)(100) = 90$ for the samples of size 10.

Therefore, s_n^2 is a biased estimator for σ^2. It systematically underestimates the true population variance. In deciding which estimator to use between s^2 and s_n^2, we will choose the one that is less biased, and so we divide by $n - 1$ instead of n.

WHAT IF WE KNOW THE POPULATION MEAN?

There appears to be a paradox here. If σ^2 is calculated by dividing by N, why does the better estimator of this value divide by $n - 1$? To answer this question, recall that in the formula for σ^2, we used the exact value of the population mean, μ. When we take a sample, we typically must estimate the population mean by using the sample mean $\bar{x}$. What if we had actually known the population μ mean in calculating a variance? Let's compare the following two estimators (Equation 1 and Equation 2) for the population variance by again constructing sampling distributions, shown in figure 11.4.

$$(1)\quad \hat{\sigma}_1^2 = \frac{\sum_{i=1}^{n}\left(x_i - \mu\right)^2}{n - 1} \qquad (2)\quad \hat{\sigma}_2^2 = \frac{\sum_{i=1}^{n}\left(x_i - \mu\right)^2}{n}$$

Now our conclusions reverse! Estimator 1 is now biased too high (by a factor of 11/10), and estimator 2 is unbiased.

The main contributor to the estimator is the "sum of squares," either

$$\sum_{i=1}^{n}\left(x_i - \bar{x}\right)^2 \text{ or } \sum_{i=1}^{n}\left(x_i - \mu\right)^2,$$

depending on whether we use the population mean or the sample means. The key is that the observations in a sample will tend to be closer to $\bar{x}$ than to μ. So when we use $\bar{x}$ in the equation, we are underestimating the variability in the data. To counterbalance this situation, we divide by a smaller number to "reinflate" the variance estimate. This idea is closely related to the discussion in chapter 10, "What Are Degrees of Freedom and How Do We Find Them?"

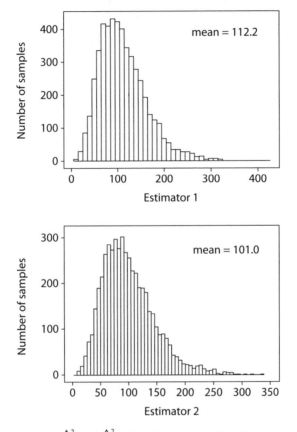

Fig. 11.4. Calculated $\hat{\sigma}_1^2$ and $\hat{\sigma}_2^2$ values for 500 samples of size $n = 10$ from a population with $\sigma^2 = 100$ and $\mu = 500$

SUMMARY

The choice of $n - 1$ in the denominator of the sample standard deviation provides a better estimator of the population variance when the population mean is unknown than if we divided by n. Here "better" is determined by having a sampling distribution for the sample variance that is centered at the population variance, that is, an unbiased estimator. If we divide by $n - 1$, we obtain an unbiased estimator for the population variance, as we have seen in figure 11.2. But if we divide by n, we have seen in figure 11.3 that the estimator is biased; it systematically underestimates the population variance. Less variability and symmetry are also nice properties for an estimator to have, but first we need to make sure we are accurately estimating the right quantity, and dividing by $n - 1$ in this instance estimates the right quantity.

REFERENCES

Burrill, Gail, Jack Burrill, Patrick Hopfensberger, and James Landwehr. *Exploring Regression*. White Plains, N.Y.: Dale Seymour Publications, Addison-Wesley Longman, 1999.

de Lange, Jan, Heleen Verhage, Gail Burrill, and Martin van Reeuwijk. *Data Visualization*. Scotts Valley, Calif.: Wings for Learning, 1992.

Freedman, David, Robert Pisani, and Roger Purves. *Statistics*. New York: W. W. Norton & Co., 1978.

Konold, Cliff, and Chris Miller. ProbSim. Software. Santa Barbara, Calif.: Intellimation, 1994.

Moore, David S., and George P. McCabe. *Introduction to the Practice of Statistics*. New York: W. H. Freeman & Co., 1989.

Rossman, Allan, and Beth L. Chance. *Workshop Statistics*. Key Curriculum, 2001.

———. *Workshop Statistics: Discovery with Data*. 2nd ed. Emeryville, Calif.: Key College Publishing, 2001. (www.rossmanchance.com/ws2/)

Scheaffer, Richard L., Mrudulla Gnanadesikan, Ann Watkins, and Jeffrey A. Witmer. *Activity-Based Statistics*. 2nd ed., revised by T. Erickson. Emeryville, Calif.: Key College Publishing, 2004.

Shaughnessy, J. Michael, Gloria Barrett, Rick Billstein, Henry A. Kranendonk, and Roxy Peck. *Navigating through Probability in Grades 9–12*. Reston, Va.: National Council of Teachers of Mathematics, 2004.

Tufte, Edward R. *The Visual Display of Quantitative Information*. Cheshire, Conn.: Graphics Press, 1983.